Kids on a Ranch

By
Ed & Jean Ann Ashurst

Foreword by
Everett Ashurst

Kids on a Ranch
Copyright 2022 Ed Ashurst
First Edition November 2022

Artwork copyright by Leann Rietmann
Instagram: leann_rietmann_art
email: desertroseleatherworks@gmail.com

ISBN: 978-1-7342951-4-6
Published by Granite Peak Productions, LLC

To order books please email or call,
azashurst@gmail.com
(520) 508-2846

All rights reserved. No part of this book may be reprinted, reproduced, stored in a retrieval system, or transmitted in any form or by any means, electronic, mechanical, photocopying, recording, or otherwise without prior written permission.

Book and Cover Design:
Cheryl Taylor, CT Communications

On the cover: Everett and Clay Ashurst
Pen and Ink by Leann Rietmann

This book is dedicated to Everett and Clay by their parents, Ed and Jean Ann Ashurst

Thank you for those wonderful years.

A special thank you to our granddaughter Leann for her artwork.

Other Books by Ed Ashurst

Non Fiction

Miracle or Coincidence:
True stories about people living on the edge.

Real Cowboys: Grand Canyon to Mexico

Wagon Boss: A True Cowboy Story

Mavericks:
Wild Cattle and Rough Country Cowboying
in the American Southwest

Alligators in the Moat:
Politics and the Mexican Border

The Life and Times of Warner Glenn

Charlie Gould: Memories of a Cowboy

Stories that Terrell Shelley Told Me

Mel Potter and Friends

Some of Them Dallied, Some Tied Hard and Fast

Fiction

Stealin' From the Neighbors

Kidnapped

Table of Contents

Foreword by Everett Ashurst vi
Chapter One 1
Chapter Two 9
Chapter Three 34
Chapter Four 49
Chapter Five 70
Chapter Six 89
Chapter Seven 93
Chapter Eight 97
Chapter Nine 102
Chapter Ten 107
Chapter Eleven 114
Chapter Twelve 116
Chapter Thirteen 125
Chapter Fourteen 134
Chapter Fifteen 142
Chapter Sixteen 149
Chapter Seventeen 158

Foreword
By Everett Ashurst

I guess the thing that is the best about being a little kid on a cow outfit is that your biggest dreams are right outside the bedroom window, and your hero is sleeping in the room next to yours. Now I know that a good dad is every little guys knight in shining armor, but when a mechanic tells a story about getting a new wiring harness installed in a 1998 Toyota Corolla, it doesn't captivate a young fella like hearing about Paul Gonzales riding the outlaw bronc off the bluff, or your dad getting the 1500-pound steer heeled and stopped right before he gored Jack's horse. I've met just as many productive, mature, respectful kids that grew up in town as I have ones in the country, but it seems like it take hobbies and sports and Dad not being at work; whereas when a boy grows up like I did, getting to go to work with your dad is your all-consuming desire.

It really is hard to communicate how bad, or even why, I wanted to get to go with the crew when they trotted out on a February morning. I remember my dad was breaking this one colt who was unbelievable hard headed. He had about 25 rides on him, I suppose, when we left one morning to gather some cattle east of the house. Seems like it was probably 30 degrees with a stiff breeze, and maybe a little snow on the ground; and about a mile above the house, we stopped and Dad got off to get a gate. After five or six of us rode through, he shut it. When he got about halfway on Apache, the horse spun away from him and proceeded to buck him off. If I had been told that morning that I couldn't go help, even if I could eat anything I wanted, play anything I wanted, or watch anything I wanted, I would have been absolutely disappointed. Not only did I want to be out there at daylight, I couldn't wait until they would let me ride a horse that would try to throw me off in the snow. And just about every kid I have ever seen who grew up on a ranch felt the same way.

One time I was doing some horse trading with a friend of mine who was the son of a famous cowpuncher. He was sitting in the doorway of his tackroom wearing a tee shirt, a pair of basketball shorts, tennis shoes, and no hat when he says to me, "You know what this is?" Not knowing, I gave him a funny look and asked what he was talking about. He indicated his appearance and said, "This." I replied that I didn't, to which he responded, "Rebellion." Grinning, I asked what he was rebelling against, and he proceeded to tell me how when he was young, his dad would whip him every time he wasn't dressed enough like a cowboy to suit the old man. The unique part about the situation was that although everyone agrees this top hand was a terrible father in almost every way, he had given his son the knowledge and experience to become very successful in the vocation of his choice, which was being a cowboy just like the man he was rebelling against.

Chapter One

Job 38: 2 - 4, Who is this that darkens my counsel with words without knowledge? Brace yourself like a man; I will question you, and you shall answer me. Where were you when I laid the earth's foundation? Tell me if you understand.

August 17, 1977

It was a year to remember. The cow market was down and had been for four years starting with the collapse of the fat cattle market in August of 1973. Cowboys drawing wages, which statistically were below the poverty line, were making more money than the ranch owners that they worked for. Elvis Presley died on August 16 leaving millions of fans in tears, and Jimmy Carter was president. Gasoline prices were high, and the economy, in general, was in the doldrums with interest rates on borrowed money reaching as high as 22 percent. But all of these facts were of no major concern to me. My world was small. The never ending specter of drought occupied my thoughts more than Jimmy being in the White House or Elvis in the grave.

 I was acutely aware that a great deal of the thousand square miles of the Babbitt Ranch had not received its share of much-needed summer rainfall. We had fed supplement feed to the cattle later than usual, going into the spring and even on into the summer. The boss was short tempered for those reasons plus others; and a great deal of the time the mood on the ranch was tense. We spent plenty of time watching clouds, hoping that they would grow and become black and heavy, and then due to their great cargo of moisture, their bowels would rupture and rain would burst forth and fall onto the dry ground causing green grass to grow. For the most part, the clouds dwindled and died and blew away without ever blessing the earth with a drop of moisture.

Jean Ann and I had only been married a short while and were living in a shack of an old ranch house at a place called Spiderweb which was the headquarters of the great historical cattle ranch that had been owned by the Babbitt family for 90 years. The first time that I ever laid eyes on Spiderweb, which was three years earlier in June of 1974, I had sworn that it was the ugliest place I had ever seen. There was a big house (by cowboy standards) that the boss, Bill Howell, lived in. You drove by it first after leaving the highway and following a dirt road for about a half mile. Then there was a cinder block building that was a shop. Next came a stick-and-stucco plastered bunkhouse, a big feed barn made of cinder blocks, and an old and shaky clapboard saddle house next to the horse corral. Attached to and continuing on from the horse corral was a big set of shipping corrals made out of cedar posts and quaking aspen rails tied horizontally to the posts. There was a shack of a house that sat next door to the bunkhouse. There had never been a drop of paint applied to any building on the place and all of these improvements blended in with the red-sandstone-badland dirt that pervaded the landscape. What grass that grew was always struggling to appear and was sparse and thin on the ground. It took a lot of acres to run a cow in that country. Nothing much had changed at Spiderweb in the three years that had passed, except for it being me and my wife that now lived in the shack by the bunkhouse.

Sometime about midday on the 17th of August, Jean Ann started talking about labor pains. I don't remember the exact time or her exact words, but she got through to me that we were going to have to go to town and have a baby. She began gathering up stuff to take along. I don't suppose it was much, maybe an over-sized purse or an overnight bag or a sack. I just remember there was stuff that amounted to preparation. My memory of her is that she was calm and focused. My memory of myself is that I was clueless as to what this all meant. I was probably thinking about riding bucking horses and roping wild cows. I was definitely not thinking of the joy of having a newborn baby or a family. I had never been one to talk about having children. I wasn't against having children, I figured they were the natural progression of life. It just happened to you. You didn't sit around and dream about being a father to a bunch of kids. A man thought about manly things: cows, horses, who was the best cowboy, or who had been left standing at the end of the last brawl.

About 2:30 in the afternoon, we started into town, and I remember driving by 89 Pasture about midway between Hanks Trading Post and Spiderweb and looking off to the west and seeing a big cloud. I watched it as we drove south toward town wondering if it would continue building and eventually turn into a rain producer. It was west of the highway about 15 or 20 miles, near the Tubs Camp or Cedar Ranch. So there we were driving toward Flagstaff, 30 miles away, and I was watching clouds and thinking about rain and drought and making a hand. Jean Ann sat next to me in the seat of a half-ton Ford pickup, and she was about to have a baby. Our first baby. I'm sure she was thinking about making a hand also—being a mother. She was always the steady one.

We arrived at the Flagstaff hospital. Jean Ann was pretty quiet. She didn't appear nervous or distraught, just quiet. I guessed she could feel things happening. It was probably about 3:30. We were directed to the place in the hospital where women had babies and where the nursery was located. A nurse took Jean Ann, and they disappeared behind some swinging doors and headed down a hallway and left me in a lonely waiting room that had several brown naugahyde-covered chairs with steel arm rests, a sofa, and a few *Better Homes and Gardens* and *Redbook* magazines. I sat down and picked up a *Redbook*. I don't remember if I had been invited by the nurse to accompany my wife into the delivery room or not, but it had already been announced by me that I wasn't going to take part in that procedure. That was no place for a cowboy.

Sitting in the waiting room was boring. *Redbook* and *Better Homes and Gardens* were boring. I sat and wondered if that big cloud had grown into a real turd-floater, a frog-strangler. I doubted it. I took a nap. Around 4:30, Jean Ann showed back up and said that we could leave. She repeated to me the instructions she had been given, "The nurse told me to go home because it was not quite time. I told her that we lived a long way from town and my doctor had said to come in early. So she said we should go out to eat or go to a movie." So the general idea was to just hang out for a while and perhaps this baby might decide to make an appearance.

We drove downtown, and I parked across the street from the Monte Vista Hotel and the door going into the Monte Vista Lounge, a favorite watering hole for all Babbitt Ranch cowboys. I got out of the pickup and walked catty-corner, toward the west, and entered the old Babbitt department store and walked up the stairs to the ranch office and got our mail. I hurried along, not wanting to abandon Jean Ann for any great length of time, although she was acting calm and not worried about anything.

I walked out of the Babbitt building and was hurrying back across the street, when just before I reached our pickup, I heard a voice calling my name, "Hey, Ed. Wait up." I looked south toward Santa Fe Avenue and the 66 Club and saw two cowboys coming my way. I could tell by their gait that they had been in the 66 Club. "Come on, let's go to the Monte Vista and have a drink." It was my good friend Jim Dolan accompanied by Paul Gonzales; both of whom were also employed at the Babbitt Ranch. They were partying and very happy to see their old friend Ed.

"No, I've got to go. I've got Jean Ann in the pickup, and, we've got to go." I had almost reached our truck.

"Oh, come on. Jean Ann doesn't care if you have a beer with us. Just one."

"No, I need to go."

"Oh bologna, you ain't in that big of a hurry! Just one drink with your old buddies, Jim and Pauly. Come on." Jim was grinning that Dolan grin that is so famous. Paul Gonzales just stood weaving and grinning, being in that zone where the perfect level of buzz was being experienced. Jean Ann gave me a look.

"No, really, I need to go." Actually, I was thinking to myself that going and having one beer shouldn't be that bad. I looked at Jean Ann. She was gritting her teeth and looking at me with a look that I had never seen before. *I guess I better not have a beer with my friends*, I said to myself. I got ahold of the door handle and started opening the driver's side door of the pickup. Jim Dolan and Paul Gonzales started walking across the street and looking at me over their shoulders and laughing.

I opened the door, and then I saw it. I thought, *Jiminy Crickets what the heck is going?* There was a watery bloody mess all over the pickup seat. It had overflowed out onto the rubber floor mats. My eyes got big.

"My water broke. We need to go back to the hospital." Jean Ann said matter of factly. She wasn't happy, but she was matter of fact.

Can babies live through the water being broke? Is this normal? Sure made a mess all over my pickup!

We raced back up the hill to the Flagstaff Medical Center, as the hospital was called, and we got ourselves back into the maternity ward. Jean Ann disappeared down the same hallway and behind the same swinging doors; and I was left with the same *Better Homes and Gardens* and *Redbook* magazines. I was by myself. Again.

I had not been in the maternity ward waiting room long enough to fall asleep when all of a sudden I heard a scream and other outrageous noises coming from the entry area of the hospital. Some woman from down that direction was screaming bloody murder. Loud, outrageous, verbalizations; the kind that would come from a female pioneer who was being scalped alive by a Comanche Indian. I straightened up in my naugahyde-covered chair and watched as the swinging doors at the entry end of the waiting room banged open, and the feet of a wailing women in a wheelchair kicked the swinging doors again, and they all entered. "AAAAAAAAHHHHHH!" The wailing woman screamed loud enough to wake the dead. The wheelchair that she sat in was being pushed by a man wearing black-rimmed glasses. I think he was wearing a narrow black tie. He looked like a stenographer, or an accountant, or possibly a librarian. I could tell that he was a mild soul, a gentle person that would have nothing in common with a Babbitt cowboy. As the dutiful man with the black glasses pushed the wheelchair past me, the woman ushered forth another scream that could have called up the witch of Endor out of her grave.

Following behind the dutiful man with the narrow black tie were the perfect duplicate of the Peanuts cartoon characters. Three children, two boys and a girl: Charlie Brown, Lucy and Linus. I swear the third and smallest child was even dragging a blanket. They calmly and obediently followed along behind the man with the black-rimmed glasses who pushed the wheelchair in which the screaming woman sat. They had all been through this drill before. The screaming woman was not in a state of distress, she was just screaming; loud, gusty screams that had purpose. The man was not agitated; instead, he was just pushing his screaming wife toward the bowels of a hospital maternity ward. The children who followed single file were quite happy to be helping in the family activity. The eldest had obviously been through this enough times to be quite well rehearsed. Linus brought up the rear sucking his thumb and dragging his blanket.

The screaming woman kicked the swinging door open that led down the hallway to the delivery room that I supposed my wife occupied. I wondered if they had several. Linus and his blanket had not passed through this barrier more than ten seconds prior to an unearthly wail ushered forth out of the woman whose husband pushed her. I could hear much commotion only a few short feet away from my place on the naugahyde couch and could tell much was taking place on the other side of the swinging doors. I wanted to open the swinging doors and peer down the hallway, but I didn't possess the courage. I'll swear that only seconds had passed since Linus and his blanket disappeared when I heard one violent wale break forth like a much-amplified strum on a rock and roll guitar; and then, I swear, there was an audible sigh of relief, then a moment of silence, then a sound like a flyswatter swatting a fly on the kitchen counter. "WWWaaahhhhh." Like the bleating of a sheep came the cry of a new-born, bloody baby!

The flyswatter sound was some doctor or nurse swatting Linus' little brother's behind as he held him upside down by the ankles, just like in the movies. And it had happened just a few feet away, right out there in the hallway! These folks were real pros. They knew how to get things done, and if I had not been such a coward I could have opened the swinging doors and witnessed the whole thing myself, but I was way too much of a prude! The woman had her fourth child right out there in the hallway. I supposed she was whisked off to a private room where her newborn, and recently spanked child, was applied to her awaiting breast while the man with the thin black tie and Charlie Brown, Lucy and Linus watched the natural progression of life take place. They were all very well adjusted. Much more than me. I never saw any of those folks again.

When the sun went down, the waiting room with the naugahyde chairs got dark. I remember thinking about it, and I decided it must be against hospital rules to have light in the waiting room after sundown. After awhile I drifted off to sleep in an uncomfortable position on the naugahyde couch, which resulted in me having a kinked neck for a week or so. It never occurred to me that there might be a light switch somewhere.

Around 9:30, I was awakened by a nurse who said to me, "Congratulations, Mr. Ashurst, you have a new son. There were some complications, the placenta broke loose too early and your son lost a considerable amount of blood, but he's going to be fine. You can come and see your wife now if you want." I looked at her with a questioning expression, but I didn't know what questions to ask. *Placenta broken?* I thought, *What is the meaning of that? Is my kid going to be okay?* The nurse was obviously waiting for me to get my act together and follow her down a hallway to where Jean Ann was. *Was she alright?*

The nurse directed me into a room, and there was Jean Ann lying on a bed on the left just after I passed through the door and entered the room. She looked exhausted. Her hair looked as if she had been picking cotton out in the sun on the hottest day of the year. She smiled faintly.

Jean Ann explained what had happened when our son was born. She said the regular doctor, whom she had been seeing once a month for six or seven months, was gone on vacation, so another doctor had filled in for him and delivered our baby. She felt that having this new doctor had been a good thing. The other doctor, who was an old guy and not too up-to-date, maybe, would not have handled the complications as well as this younger doctor who seemed to be really on the ball. It was all gibberish to me. I was glad the kid was alive.

After a few moments another nurse came into the room and said, "You can come and see your new son now, Mr. Ashurst."

I got up out of my chair and followed the nurse down a hallway. We came to a place where the wall was made up of huge plate-glass windows. The nurse pointed into the room on the other side of the windows. There were about a dozen incubators scattered around the room, most of which looked empty. There were several nurses in the room taking care of business. The nurse who had brought me down the hallway was still standing next to me, and she said, "There, that's your son, right there." She pointed to the incubator closest to me, just several feet away but on the other side of the big window. I looked and saw this little child, about the size of my fist. He had tubes stuck in his nose and hoses taped to his chest and arms, with IV needles attached to the hoses and the ends connected to needles that were stuck into his arms. It appeared to me that he needed an emergency room more than an incubator. I stared down at the helpless child and marveled at how tiny he was. I wondered at the hoses and tubes they had stuck into him. I noticed that none of the several other babies in the room had hoses or tubes applied to them.

Suddenly I felt sick to my stomach. I thought I was going to faint. I put the palms of my hands against the plate-glass window and leaned into it, bracing myself because I thought I was going to faint and fall down. I began to shake. I felt dizzy. I thought I might vomit. I leaned against the wall for a moment trying to collect myself. The nurse had left, and I was alone and glad of it. I didn't want some woman to know I was sick—for no reason. I looked at my son. Everett Gene Ashurst, that is what we had decided we were going to name him.

I thought about going into the room and telling the nurses they needed to check on him. They were only several feet away from his incubator. I decided that perhaps they knew what they were doing. Perhaps. My legs came back to life under me, and my strength returned, so I turned and walked down to the room where Jean Ann was. I sat down in a chair and looked at her. I felt guilty, like we had been on the Titanic as it was sinking, and I had pushed her overboard, and then someone else had rescued her by pulling her into a lifeboat. She looked like she had experienced a Titanic event, and I had slept through the whole thing in the dark waiting room. I felt very small. I didn't know that life was never going to be the same again.

Chapter Two

I first went to work on the Babbitt Ranch in May of 1974. I believe I showed up on the 11th of May, although I'm not sure of the exact date. Bill Howell, the ranch manager or wagon boss, whichever title you prefer, was cooking for the crew because he had not been successful at hiring a cook. Upon arriving at the place known as Cedar Ranch, where the crew was camped, Bill offered me the cooking job, which I declined. After receiving my negative reply about the position of cook, he said that he had one mount of horses left that were the dregs of the remuda, and he would give them to me if I wanted them. That offer I accepted. The cowboy crew that I now joined consisted of not only Bill, but his brother Harvey commonly known as Boog, Ken James, Bud Watson, Gordon Mecklenburg, Cotton Elliott, Bill Raye, Tom Jones, and I made nine. Several days later Bill hired a drunk cook by the name of Lee who was a fair cook and quite comical, but not very dependable.

The first day I worked for Bill Howell, I rode a horse by the name of Pole Cat. He bucked all day, but I was never unseated, although I came close several times. We gathered a herd of cows and calves and trailed them to a waterhole called Mesa Butte Storage and branded the calves, which amounted to a total of about 120.

This was the day that Lee, the new cook, showed up. This turned Bill loose from the cooking chores, so he showed up at Mesa Butte Storage in a pickup with all the necessary items needed to brand calves: branding irons, vaccine, etc. He also had his three sons with him, Vic, Tim and Tom. They were about 12, 11, and 10 years of age. They wore cowboy clothes, including cowboy hats, and even though they were little kids, they looked the part. The boys also knew how to help. They could brand a calf or vaccinate it, and help hold the small ones down on the ground while it was branded. There was no little-kid nonsense going on. The three boys were all business and were making a hand. It was obvious to me that their father had spent a lot of time schooling on them. I noticed that he kept his eye on them as they maneuvered through and around the nine grown men that were present, but he said very little, if anything, to them. I was impressed by their behavior. They acted like little men.

I had heard a considerable amount about Bill Howell before I went to work for him. I had been told that he was a top cowboy and cowman and that he was an exceptional bronc rider. I had heard that when he took over as the Babbitt Ranch manager, only five years before I came, that he had been dealt a considerable amount of hostility from some old Babbitt hands who resented him being promoted to the position. Some nasty rumors had circulated around Northern Arizona saying he was incapable of handling the job. One cowboy who had previously worked for Babbitts, but not currently, who was a drinker and a noted tough guy, drove to Spiderweb a few days after Bill took over and threatened to whip him in a fist fight in the bunkhouse, but he had failed to get that feat accomplished.

By the time I showed up there, Bill was well established as the man who ran the outfit. I was impressed right away with his adherence to duty. He was a company man and held his employers, especially old John Babbitt, whom he answered to, in high regard. He only said good things about John, and I definitely got the feeling that nobody on the crew had better say anything bad about John Babbitt; but I never saw that premise tested, because John was respected by everyone.

I worked hard at being a good hand and was confident that Bill liked me. I had come to work there at the perfect time in my life. Had I went to work there several years earlier, my cowboy skills would not have been honed into journeyman status, or had I showed up several years later, Bill might have been too preoccupied with developing problems he faced to appreciate my skills, but at this point in Bill's life, he was still young, 39, full of zest and a love of cowboy work. He liked what he did for a living, and he enjoyed working with men who liked the same thing. I loved it, and Bill and I hit it off very well. He was an inspiration to me. I became aware of a seemingly different modus operandi, or perhaps even worldview, that Bill Howell exhibited compared to that which I operated under. He obviously thought that working hard and being good at what you did was a desirable goal. But even more than that, he had a steadiness, and seemed determined to be successful as a result of taking care of business, and an open disdain for frivolity and mediocrity. He was a stayer.

I was not a stayer. I had never been a stayer. I was raised by parents who, at their 50th wedding anniversary, counted up a total of 57 moves in 50 years of marriage. I had changed schools 12 times from 1st grade to the 12th grade. I was a journeyman member of the elite cowboy subculture that considered changing jobs on a regular basis (several times a year), drinking lots of alcohol, and getting in barroom fights activities both to be admired and coveted. I slowly began to recognize in Bill Howell a difference in philosophy than what I had seen in a host of the cowboys I admired. I began to see that he had a better way, and as a result I put some roots down into the Babbitt Ranch.

Jean Ann and I started going together on Easter Sunday 1976. We danced to a couple songs playing on the jukebox in the bar at Hank's Trading Post after a jackpot team roping held in an arena close by. I think I had even won some money that day. She and I have been an item ever since.

The next winter, things got a little sticky on the Babbitt Ranch, or, perhaps, we should say a little ugly. One of the top men on the payroll, someone who had a considerable amount of responsibility, got fired as a result of some malfeasance on his part; so Bill gave his brother Harvey Howell a big promotion so he could fill the position the other man vacated. In the midst of this, Jean Ann and I were planning on getting married, and, at the same time, Bill's marriage to his wife Gloria was coming to an end. And then a few days before we were to get married, Harvey Howell quit the outfit to take what he thought was a better job, even though Bill had, only a month before, gone to John Babbitt and asked him for a big promotion and raise in pay for Harvey. On top of all that, it would not rain. Bill Howell was about as happy as a rattlesnake who was being taunted by a cowboy with a shovel. That was the situation on the ranch when Jean Ann and I got married.

I remember when we were first married we drove into Flagstaff and went to the movies on a Saturday night and watched *The Shootist* starring John Wayne, Ron Howard, and Lauren Bacall. By coincidence we ran into Bill, Gloria, Vic, Tim, and Tom Howell who had driven into town to watch the movie; and so we sat next to them in the theater. That was a good time, and probably the last time Bill and Gloria went anywhere together socially. They were divorced soon afterward.

The whole mood got tenser as spring wore on into summer. The night Everett was born, I returned home to Spiderweb about midnight. Bill, Vic, Tim and Tom arrived home at the ranch only minutes before me. They had been in Flagstaff competing in a jackpot team roping, and they were unloading horses and tack when I pulled up in front of the shack where Jean Ann and I lived. Bill and the boys were only a few yards away. I walked over to where they were busy unloading their stuff; and the conversation went something like this, "Jean Ann had a baby tonight. It's a boy. We named him Everett Gene."

"Congratulations," Bill said, and the conversation was over. Jean Ann was Bill's youngest sister and was 22 years younger than him, there being eight other siblings in between the two of them. I turned and walked to the house thinking to myself that Bill's reaction would have been no different had I said something like, I just ran over and killed a rattlesnake.

Making a hand, adherence to duty, doing what was best for the company and the livestock were still at the top of the list of requirements to work at Babbitts. Worrying about whether or not your sister had made a successful delivery was somewhere below that. I understood and I determined to work at becoming the best hand that ever saddled a Babbitt horse.

Jean Ann and I settled into married life living on the big outfit. Having a child scared me because Everett seemed so fragile, so small like a piece of fine china that would easily break. I wasn't sure if they made glue that could fix a broken kid. Jean Ann was a natural mother. She could have been a pioneer. She liked living on a ranch a long way from town. She didn't need to go to town all the time to shop. She bought groceries once a month and got the mail at the same. It didn't matter because we didn't get much mail anyway. She used cloth diapers and never acted like she had planned on anything else, so we weren't running to a store to buy disposable diapers. Jean Ann didn't need disposable diapers to be happy. She never had an identity crisis.

Pat Lauderdale and his wife Sherry had a baby girl about Everett's age named Johnnie. Once in awhile they would stop by. We would visit, and Johnnie and Everett would try and play, but they were pretty small and mostly the two of them just crawled around.

Chapter Two

There was a cowboy, a confirmed bachelor, living next door to us in the Spiderweb bunkhouse whose name was Bill Miller. Bill and I were good friends. He had a big walrus mustache and was a very soft-spoken, almost shy, individual. He was always very polite around women but would remove himself from their presence if given the opportunity. He was a loner. Several times Bill came over to the house and would visit for a little while and drink a cup of coffee, but he never stayed long. So one cold winter day when everyone wanted to be inside by the stove if possible, Bill Miller came over for a visit. Jean Ann made him a cup of hot coffee, and we sat for a while talking. Bill held the hot coffee in his hands and then suddenly it happened; it just sort of slipped out by accident. Bill passed wind. He let a fart escape that he didn't intend to turn loose. I suppose my eyes looked the other way, or maybe I laughed a little, I don't remember. I think Jean Ann looked the other way trying to keep from embarrassing the poor guy. But the noise and everything that came with it was more than obvious. Bill got so uptight he squeezed the porcelain coffee cup handle and broke it, so there he was with the handle in one hand and the cup in the other. He stood up and walked out the door and never came back.

An old bachelor named Mike Lenton was the waterman on the CO Bar at that time. Mike had worked for Babbitts for 40 or 50 years and had been faithful to the company through the years of World War II when help was hard to find because all the young men were gone fighting Nazis or the Japanese. The word was old John Babbitt wouldn't let any foreman, no matter how powerful he thought he might be, fire Mike regardless of how lazy he might be. Mike had stayed with the company when help was hard to find and John Babbitt was going to stay with him.

Mike was famous for never taking a bath, and only one time in 50 years had anyone heard the shower run when he was in the bathroom in the Spiderweb bunkhouse. That date was marked on the calendar by Raymond Holt who was just as old but cleverer. But, truthfully, it was not known for sure if Mike actually got under the water. But Mike was always clean shaven, and he did not smell bad. He had nice clothes and a good hat, but he wore that only on special occasions. On many days he would find an excuse to drive into Flagstaff to buy a pipefitting, or a spark plug for a generator; and he would park his company pickup in front of the Monte Vista Lounge and go in and give the town folks an update on how much water was in Big Boy, a 500,000-gallon water storage that was on the pipeline running from Slate Well out into the Double Knobs Pasture. All the cocktail-drinking crowd at the Monte Vista knew about Big Boy, and they would laugh and ask the other Babbitt cowboys who came into the Monte Vista to have a beer how much water was in Big Boy.

Mike had one eye that was good and the other had been ruined by a grain stubble that poked it when he fell down as a small child running through a recently cut grain field in North Dakota where he had grown up. His bad eye and greasy work hat were as famous as Big Boy. Mike was a character and a permanent fixture on the Babbitt Ranch.

That spring of 1977, the first we were married, was exceptionally dry. Bill Howell did some things that year that were different than most years, trying to adjust to the drought conditions. One of those changes was to brand calves in places where we normally wouldn't because there were no corral facilities, but there was feed and water. One of those places was Coyote Storage in 89 Pasture. Bill began roping that day, catching the calves around the neck and dragging them to the branding fire bouncing out on the end of his rope, while four or five men held the herd horseback and several of us worked on the ground, flanking, branding, castrating, et cetera. I was part of the ground crew. Bill was in a particularly nasty mood that day. That was about the exact time that Gloria left and moved to town, and Bill had become a bachelor. It was dry and hot and the cattle didn't look good. Bill was probably the best roper and all around horsemen I've ever known, but he also had a temper. He was riding a colt that he had raised that he called Smokey; and Smokey, the horse, wasn't working good enough to suit Bill, and nothing in general was going well. It was man against beast.

Smokey never saw the day that he could have bucked Bill off, but the horse did something worse; in an attempt to give himself some escape from Bill's training, the horse reared up and fell over backwards and came down hard right on top of Bill who was crushed under the saddle. It was ugly; Bill laid there in obvious pain. He wasn't what you could call a gentle patient. I walked out and spoke to him with measured words. I didn't want to say the wrong thing. With limited communication I learned that he needed a little help getting up, so I got several men from the ground crew, and we helped him walk to the pickup that had been used to bring all of the branding equipment to the branding site. He instructed us to help him up into the bed of the pickup and make a space for him to lay down on the steel floor. So we did what we were told. Then he told me to finish roping the calves and carry on with the work, which we did while he lay down on the steel bed gritting his teeth. One of the men on the crew, a fellow named Butch, said to me, "I think we better go call an ambulance. Bill looks really bad."

We could see Hank's Trading Post and Highway 89 five or six miles off to the east. I told Butch, "If you want to call an ambulance, get on your horse and lope down there to Hank's and call one. I am not having any part of it." I knew that Bill was hurt bad, but I also knew that he was awake and in full possession of his faculties. If an ambulance had showed up there to get him, he would've cussed the ambulance driver out and sent the ambulance back to town empty; and the man who had called the ambulance would've got a cussing he would never forget.

Years later Bill had some x-rays taken due to a different medical problem, and the doctor asked him, "When did you break your back?" Bill replied that he had never broken his back. And the doctor quickly came back with, "Yes you have," and then he showed him the x-rays that revealed scars of a fractured vertebra. Bill was so lean and his muscles so hard that the fractured spine never moved, and he just gutted it out and worked in pain but never went to the doctor. That's the kind of family I married into.

Jean Ann and her family, the Howells, were hard-working northern ranch people. Her father and mother had moved to southeastern Montana in the late 40s but were originally from the Sandhills of Nebraska. Her father was an excellent horseman and cowman who hated working for other people but preferred scraping by at the poverty level working for himself over taking orders from someone else. When she was a child, the family went through several hard years when things got bad enough financially that they burned worn out tires in the furnace of their house instead of coal, and the resulting black smoke coming out of the chimney advertised the situation.

For several years, Jim, her father, traveled the Midwest working as field representative buying Angus cattle for Jim Madden, owner of the livestock auction at St. Onge, South Dakota. Jean Ann went with her parents on a few of their trips to Missouri and Arkansas and remembers him stopping by places with Angus cattle and trying to talk the owner into shipping them to St. Onge to sell.

If a man's cattle were poor, and the place falling down, but the man chose to keep his cattle, Jim would be angry at the man's stupidity and as they drove away, Jim would be calling the man profanities, and Virginia would be shushing Jim for using curse words. Jean Ann was thinking that her dad didn't have any business being angry at the man; they were his cattle, and it was his choice.

In Ekalaka, Montana, Jim went into business with a man who claimed to have experience in the car business .His partner in the used car lot skipped town with the partnership checkbook, leaving Jim with a large debt in a business he knew little about. This piece of bad luck created more hardship, but Jim Howell took it like a man and never spoke ill of the crook who had left town with the bankroll. In the mid-60s, he got a break and was offered a partnership on a ranch a few miles south of the Black Hills of South Dakota, and so when Jean Ann was 10 years old they moved to a ranch near Smithwick, South Dakota.

Jean Ann's mother, Virginia, came from a family of successful Sandhills ranchers from Mullen, Nebraska. Like Jean Ann, Virginia was country when country wasn't cool. She raised 10 children through good times and bad, always keeping a clean house and a well set table, serving food that was homemade. Her cooking was famous even in a time when lots of women could and would cook. During the years after the used-car-lot catastrophe, they ate lots of beans and homemade bread; but it was always good homemade beans and bread, and they ate it in the house heated by discarded used tires. Virginia was a committed Christian. She always went to church, and if available she went to the holy-roller variety. There wasn't a lot to choose from, the dead, boring Baptists or the Pentecostals where prophecy and speaking in tongues and lifting your hands during the service was commonplace.

I had been raised in a similar atmosphere. My mother was a committed Pentecostal Christian. Like Jim Howell, my father didn't go to church much, but my mother did. Like Jim Howell, my father loved cows and horses; but my father wasn't much of a businessman, and he never stayed put anywhere for very long. They were both honest and hard-working, but they both possessed a strong temper and an aversion to being bossed by someone else.

In one respect Jim Howell and Henry Ashurst were very different. Jim was a lifelong conservative Republican, whereas Henry was a dyed-in-the-wool Democrat. My parents thought

Chapter Two

Franklin Roosevelt was the savior who lifted America out of the depths of the Great Depression. My father's uncle and namesake, Henry Fountain Ashurst, was one of Arizona's first United States senators. Arizona's greatest son, Barry Goldwater, wrote a book about my famous great uncle Henry. My grandfather Edward Bates Ashurst was an attorney and had been a superior court judge for eight years. In 1939 he ran for congress but lost. He was a very fine man who I knew well. He was honest. He was a Democrat.

Ed took me with him down to visit his folks one weekend after we had been dating for a few months. I thought it meant he was kind of serious about our relationship. I learned later that when he wanted to visit his folks he just took whatever girl was his girlfriend at the time. Multi-tasking.

On the second day of the visit, Ed's dad, Hank, started quizzing me down about my family's political beliefs. It never crossed my mind that Ed's folks could be anything but Republicans, so I was answering his questions as if talking to a person of common opinions. I was very adamant about my Republican convictions as were my parents. When President John Kennedy got shot, everyone at school was all shook up, and my friend's mother was crying like her father had died. When I got home, my mom was praising God that we had been delivered as a nation from that evil man who had got himself elected on money made from bootlegging alcohol. I looked at my mom and thought—Hmm, this is one of those things you don't tell at school.

Our teacher decided that we kids needed to have the opportunity to express our feelings, so she began questioning us about how we felt. I was, of course, not talking, but neither were the other seven-year-olds. For some reason the teacher/counselor zeroed in on me to make me open up. Finally I told her I didn't know why everyone was making such a big deal about it. Our country wasn't going to fall apart because we had a form of government in place with a vice-president to take over. And besides that all the people had not voted for John Kennedy. He had won by only a very small margin. I suppose this surprised the woman, but she moved on and asked me about his wife, wasn't it sad that she had lost her husband. I informed her that we didn't even know, for sure, if she loved him. She might just act like she loved him because it was her job to act like she loved him.

Well, then, what about his kids? I replied with passion to that question, "Yes, what about his kids? Everyone is going around worrying about themselves, and those kids will never have another dad. I got really sad as I said that, and all the other kids must have too, because the counselor drilled in on that emotion and told us all to go home and tell our parents that we wanted to do something nice for President Kennedy's kids.

I don't know how Dad found out about this, but after supper, he, with all the skill of a man dealing with his 10th child, led me to tell the story of what had happened in school that day. Mom was horrified at my theory of the president's wife just faking affection, but Dad settled her down and moved the story on to the end where once again I got very sad thinking about those kids that were never going to have another dad. Dad might have known how to pull a story out of me, but my mom knew what to do with problems too big for people. She said, "Do you want to pray for those kids?" Yes I did.

After a few days, the teacher asked all of us second graders what we and our parents had talked about and done concerning the Kennedy kids. Some had sent cards, and I don't remember what all, but I knew in my heart that my mom and dad and I had done the very best thing that could be done for those kids.

I suppose that experience made Democrats' kids seem like regular people, but when the kids grew up, I knew they must have lost their minds to think big government was a good thing.

So that was the 20-year-old who sat visiting with her boyfriend's parents in their kitchen, talking politics. After a little conversation, Hank said that they were Democrats. I am embarrassed to admit I was as spooked as if I had found out that I was staying in the home of practicing polygamists. The thought flashed through my mind—I don't even have my own vehicle to get away in—and right at that moment, Ed walked into the kitchen.

Ed, the Ed I had known for over a year, the Ed my brothers and nephews liked, the Ed that had always acted and talked normally. Was this relationship about to be over? "Are you a Democrat?" I asked him with big wide eyes and my face a few shades paler than normal.

I had no idea that to say what he did in front of his parents was a very brave thing, but he answered calm and steady, "Well, I voted for a Republican in the last election."

When Jean Ann and I got married she continued going to the Assembly of God church in Flagstaff. She didn't go often as it was about a 40 mile trip to get there, but she did go, maybe several times a month. She wanted me to go but I wouldn't. I didn't want anything to do with church. Especially an Assembly of God church. I had grown up in Assembly of God churches, that being my mother's favorite flavor of Pentecostalism when I was a child. If blindfolded I could recognize an Assembly of God church just by the smell. I swear that is the truth. Jean Ann wanted to pay tithes to the Flagstaff Assembly of God church. Ten percent of my $450 dollars a month Babbitt Ranch wages. I finally, begrudgingly, allowed her to pay five percent; her half of ten percent. That was our agreement.

In those days on the CO Bar, several of the camp men would spend the summers and winters in different camps. When I first went to work at Babbitts, old Raymond Holt lived up at Wild Bill Camp at 8000-foot in elevation in the summer and down at the

River camp, on the banks of the Little Colorado east of Savage well, in the winter months. The fact that Raymond had to move twice a year was of no consequence to him because he was a lifelong bachelor, and what few possessions he had were easily packed up and moved. Married cowboys weren't given much more time to move than bachelors like Raymond. Bill would look at you someday at noon and tell you he would give you the afternoon off so you could get you and your family moved. For Jean Ann and I that meant moving from Spiderweb to Cedar Ranch, 26 miles away. I have no memory of devoting more than an afternoon to help with the move. I was much too busy trying to be the King of the Cowboys or Clint Eastwood's rival to help my wife with mundane duties like moving chairs, trunks, or diaper pails. Barbara Mandrell thought she knew what she was talking about when she belted out, "I was country when country wasn't cool," but Barbara Mandrell was never married to a cowboy with a two-watt brain and a Clint Eastwood ego. Neither was Gretchen Wilson who became famous singing Redneck Woman. Jean Ann was, and still is, more country than either Barbara or Gretchen.

On one of the rare nights when Ed came home from the wagon to shower, sleep, get up and leave at three in the morning, he told me that Bill would be giving him time to come help me move to Cedar Ranch some day soon, so for me to be ready. I had moved six or seven times between leaving the home of my parents at almost 18 until moving in with my husband when I was 20, so I knew how to move. It was just something about how he said "help" that caused me to think that meant he would help from the getting of the boxes to the putting away of the dishes. Ed had told me that he had to change schools 12 times in his 12 years of schooling because of moving with his parents, so I figured his help would be real good help. Actually, I just couldn't wrap my mind around getting ready for an unknown moving date, so I didn't.

Ed showed up late one evening, much to my happiness, and looked around at the house with everything still in its place. "Bill gave me time to help you move tomorrow. Are you taking this furniture and stuff?"

"Oh, sure. We have to have something to sit on up there."

"Are things packed?" He could tell that they were not. "You got boxes?"

"No, I haven't collected any boxes yet," I replied cheerfully.

Ed was showing signs of being stressed. "We have to move tomorrow. That's the day Bill gave me to move. How are we going to move without boxes?"

"We could go to town and get boxes."

"No, we cannot go to town and get boxes and get moved and get me back in time to rope horses tomorrow night. Why didn't you get boxes?"

"I don't know; I guess I thought you were going to help."

"I'm here to help you load your stuff so you can have it at your summer camp. Do you have anything to pack this stuff up in?"

I walked around the house looking for some kind of containers with which to move a household. I saw the paper grocery sacks. There were lots of them, saved there in the pantry, because they were handy for so many things, and I had been well trained by my folks to save handy things. They saved coffee cans because they were useful for so many things.

The place where we lived in South Dakota had a one-car garage with an attic over it. When we moved there, when I was 10, the folks decided to store the empty coffee cans in the attic over the one-car garage. It was pretty easy. When a few cans collected, a couple of us girls would carry them to the garage, open the car-sized door, because that was the only door; one girl would climb the short ladder and hold on with one hand while accepting a can from her partner with the other and throw it as far as it would go.

Dad drank a lot of MJB coffee, three-pound cans. One day when I was making Dad some coffee and telling my folks a story, I got distracted and smashed the can's top, which was the required procedure on all cans so if a cow walked too close to the dump, she would not be able to step in a can and get it stuck around her hoof and cripple her. In my distracted state, I smashed the top of the coffee can and threw it in the trash. My folks looked at me as if I had thrown away the butter and kept the paper. I looked at their faces and wondered about the attic full of handy coffee cans that had yet to fill their purpose in life. Oh, well, I thought, one day I will be on my own, and I will not do such impractical things as save a whole garage attic full of coffee cans.

Now I was on my own, and those saved-because-they-are-so-handy paper sacks were the only thing in the house to collect a few items together in. I pulled them off the shelf and carried them out to the big room and announced to Ed that I had paper sacks to move in.

Ed was not impressed and declared we would deal with it in the morning. To bed we went.

Morning did not make things look better to Ed. I was happy to have him home with me. He was very petulant. I commented that he was acting grumpy. He replied that he hated moving and was not going to be happy until we were moved in at Cedar Ranch. He announced this as he smashed things into the paper sacks.

In a few hours we got everything stuffed into a place, loaded Everett, and headed out. The short way to Cedar Ranch was across the ranch on dirt roads, about 26 miles; through town on mostly pavement was 75. It was very dry, and the road through the ranch was powder. I meekly asked if we might go through town to save the furniture and things that were in the open-topped trailer from being permeated with dust.

NO! Only high-maintenance wives needed something as demanding and time-consuming as that. High-maintenance wives were an embarrassment to their husbands.

I was so excited to be moving to Cedar Ranch where I would have more than four feet of kitchen counter top on which to prepare a meal, the elevation was higher, *the view beautiful, Ed would be home most every night all the rest of the spring works, and the horse corral sat right out my front door where I could watch the cowboys with their horses every day. What was a few days washing and vacuuming every piece of our household and personal items compared with that reward.*

The summer of 1978 Jean Ann and I lived at Cedar Ranch in the big house down close to the barn and horse corral. The truth is Cedar Ranch is as much a home to me as any place I've ever lived and is as beautiful a place as I've ever known. Those were good days. I broke some colts that summer and took care of the Slate Lake Pasture. Everett was crawling around on the floor of the house, and Jean Ann was pregnant.

In those days Mormon Lake, 30 miles south of Flagstaff, had a roping arena where a big jackpot team roping was held every year on the Fourth of July weekend and Labor Day weekend. Those were probably the biggest ropings held in America at that time, as far as number of teams entered was concerned. The best ropers in the world would show up and compete there.

Jean Ann's brother Harvey had quit Babbitts at the same time Jean Ann and I got married and had taken a job running the Apache Maid Ranch that ran from the mountain just a little south of Mormon Lake and on south to the Verde Valley. In late August 1978, after working for Oscar for about a year and a half, Oscar fired Harvey for no apparent reason other than he got out of bed on the wrong side that morning. Harvey was doing a fine job. This happened right before Labor Day weekend.

The Verde Valley Rangers Sheriff's Posse were the producers of the Mormon Lake roping events, and that year they decided they were going to have a special roping during their Labor Day event, which ran for three days. The special roping was going to be a five steer average for $50 dollars a team, which was considered a huge entry fee. They were going to limit the roping to 50 teams. Harvey and I decided to enter. I believe this decision had been made before Oscar fired Harvey leaving him unemployed, but, in spite of that, we entered anyway. Harvey stood in a line all night so he could be sure to get us entered. There was actually that much interest in the roping, making people stay in line all night, assuring them an opportunity to get entered.

This was before the advent of computers and electronic entering. Everything was done by hand. Some of the best ropers in the world were entered including Gary Mouw, Walt Woodard, and several NFR contestants. Harvey and I won first in the average, roping five steers in 60 seconds and pocketed $450 apiece. Harvey roped both hind feet on all five steers, which in those days was a big deal. Harvey was temporarily unemployed and roped like his life depended on it. Four hundred fifty dollars was a month's wages to me, and I had a pregnant wife. But unlike Harvey, I had a job.

That fall of 1978 the Babbitt Ranch had an excellent crew of cowboys. Pat Lauderdale held court as a self-appointed expert and top man, as well as the leading old-timer and teller of stories (he was probably 41 years old). Pat was opinionated and always picking on some poor soul who was at the bottom of the pecking order; but he was a good cowboy, and as long as he wasn't mad at you he was fun to be around. I was not intimidated by him in the least and considered myself to be his equal at any part of the cowboy craft. I enjoyed Pat.

Pat was not a complainer, and no matter how hot, cold, or miserable the weather, he enjoyed working cattle and was making a hand even though he might be cussing someone who had the misfortune of falling out of his favor, which happened frequently. There were two brothers there from Gila Bend named Paul and Ruben Gonzalez, both of whom had learned to punch cows under Pat's tutelage. They were exceptional cowboys, could ride anything and could really rope. They hadn't a care in the world. Cisco Scott was there, and had been off and on for several years. Cisco would've been about 22 years old. He was famously quiet, sometimes not speaking for several days but was also a good hand. Bill Van Pragg was there and his wife, Dorothy, was cooking for the roundup crew and was as good a cook as ever lived. Bill and Dorothy had stood up for Jean Ann and me when we got married and they were good friends. Bill Howell was always very fond of Bill and Dorothy.

Around the first of October, we camped at Kendrick Park at 8000-foot elevation, and we gathered the Wild Bill Allotment. This took about five days. While we were camped there a famous photographer who had a studio in Flagstaff whose his name was John Running came out to Kendrick Park, arriving about suppertime. He was wanting to take pictures of cowboys, and Bill Howell gave him his blessing. Jean Ann happened to drive up to where we were camped for a visit and brought Everett along, who was now about fourteen months old or close to it. John Running took a picture of me and Everett. He also took pictures of Pat Lauderdale, Ruben Gonzalez, Cisco Scott and old Mike Lenton the waterman, and Dorothy the roundup cook. All of these photos were put into a magazine and were also exhibited in an art show. I believe that they were also put in an issue of the *Arizona Highways* magazine.

The photograph of Pat Lauderdale became very famous and has been reproduced in countless magazines, internet blogs, posters and hundreds, if not thousands, of places where a picture of a real-life cowboy was needed. In the photo Pat is standing sideways about 45 degrees looking to his left with his head cocked slightly to the right which made it face the camera. He had a black hat on complete with eagle feather, a black walrus moustache, a vest and chaps. I'm sure you have seen it, probably dozens of times. You cannot miss the cocky, self-assured smirk on his face. It is a classic, and I was there to witness its taking.

Over the course of five days, we gathered 800 cows out of the jackpine thickets and quaking aspen and threw them into the pasture that took in Kendrick Park, probably about four sections. Then after breakfast we rolled up camp and sent Dorothy down off the mountain to Cedar Ranch about 12 miles distance while we gathered the cows and prepared to move the herd toward Cedar Ranch, making it to Slate Lake the first day. There was always horses that bucked while we camped at Kendrick Park as it would be cold, around October first, and the ground was soft there around the corral where we would keep our night horses. The morning we were to leave, the horse Bill Howell was riding blew up and bucked downhill towards the edge of the big dirt tank adjacent to the corral, and he slipped and fell in the slick mud. Bill landed underneath him and momentarily hung a spur under his flank cinch. But when the horse jumped up, Bill came clean and loose, and the horse ran off, but was soon caught by someone.

Bill had been completely submerged in the icy water. Dorothy had left with all of our beds and camp paraphernalia, so there was no opportunity for Bill to change into dry clothes. He rode in wet clothes in cold temperatures as we trailed the herd down off the mountain to Slate Lake, which took all morning.

That fall we also had an old-timer on the crew by the name of Tubby Greenhaw. Tubby had worked for Babbitts a number of times over several decades, and Bill Howell like him. Tubby was no part of a good cowboy but decent enough to get by. Bill had him wrangling horses and hooding for the cook, which mainly meant chopping a little wood. Tubby was a bachelor, having had a beautiful wife at one time, but she, like him, liked to drink whiskey far more than keep house; and their marriage did not last very long.

Toward the end of October we always weaned a big bunch of calves at Aso, which was one of the biggest days of the entire roundup. We had been camped at the Tubs, but the day before we did the big weaning we moved the wagon down to Spiderweb. So the act of moving camp and gathering pairs to prepare for weaning 1200 calves, plus all of the other moving parts that needed to be dealt with, created a very big day in the midst of the busiest time of the year. In other words, Bill Howell had a lot on his mind. It was not a good time to be messing around and doing something stupid.

We rolled up our camp at the Tubs way before daylight and got Dorothy's kitchen and our beds all loaded on the chuckwagon so she could get moved without any further help. Bill sent Tubby to gather the Tubs horse pasture by himself and move the remuda to Aso, and the rest of us took off and gathered the 1200 pairs that were in SP Pasture and put them in a smaller trap at Aso. We were going to separate the cows and calves the next day. When we got the cattle gathered into the Aso corrals we headed and heeled a few big, unbranded calves and branded them. Everything was going good. Bill was happy and whoever roped while we branded roped very well, which always made Bill even happier. He really enjoyed working with good cowboys who were making a hand. Around 11 o'clock, Tubby showed up with the horses, and we threw our ropes around them and changed horses, getting a fresh one to finish the day off. About noon Dorothy showed up with the chuckwagon and paused long enough to feed us some steak and bread for lunch, along with some hot coffee she heated over a quickly made campfire. All of this had been strategically planned

Chapter Two

out by Bill Howell who was an absolute master at logistics and management of his time. Things were going good.

The plan was that after a quick lunch, eaten by the blaze of a campfire outside the Aso corrals, Dorothy would proceed down off the mountain to Spiderweb about 15 miles distance. Hooked to the old Ford truck that we called the chuckwagon was a small camp trailer that Bill and Dorothy lived in while we were camped around at various spots on the huge ranch. Dorothy was the only woman allowed to stay with the wagon, and she and Bill were married. It was all very proper. The rest of us slept in bedrolls

Tubby was supposed to follow Dorothy and the chuckwagon down to Spiderweb driving Bill Howell's new ranch pickup. I don't remember how Bill's new pickup had made its way to Aso, but it was there and needed to be moved. The remuda that Tubby had moved from the Tubs Camp to Aso was going to stay at Aso and then be trailed to Spiderweb the next day. So there they went after lunch, Dorothy driving the green Ford chuckwagon and pulling her little camp trailer with Tubby following in Bill Howell's brand-new green Ford pickup. Babbitt's never owned or used anything but Ford trucks because Jim Babbitt owned the Ford dealership in Flagstaff. The ranch trucks were always green. Everything that day was going good.

Bill Howell had given Tubby orders to follow Dorothy down to Spiderweb and help her get unloaded and her camp and kitchen set up; and then at four o'clock or so, he was supposed to return to Aso in Bill's new Ford pickup and pick us up and take us down to Spiderweb where we were now camped. While Dorothy and Tubby accomplished all of this, the cowboy crew was going to stay at Aso and do some more cow work.

As they made their way down to Spiderweb with Dorothy in the lead, they went down the pipeline road by Black Storage and hit Highway 89 near Hank's Trading Post about four miles south of the Spiderweb turn off. Dorothy, being of a good heart, decided to stop at Hank's and buy the crew a case of cold beer, after all we had been working hard for six weeks or more and had no opportunity for even a single day off, not even Sunday. She figured we deserved some refreshment so she stopped and ran into the establishment and purchased a case of Coors and walked outside and put it in the front seat of the Ford chuckwagon, which she then fired up, and headed out across the gravel parking lot toward the highway with Tubby close behind driving Bill's new pickup.

When she got to the edge of the pavement she stopped to allow an oncoming semi to pass by, but Tubby, blinded by the cloud of dust the chuckwagon and camp trailer were making, failed to see her sudden stop. He was following close, wanting to make a hand. Maybe he could smell beer.

Suddenly through the mushroom cloud of dirt and dust, the tail end of Dorothy's camp trailer appeared directly in front of Bill's new Ford pickup. The Ford pickup with less than 2000 miles on it. Tubby panicked, realizing he was about to hit Dorothy's little house on wheels, and he steered sharply to the left to avoid the collision; and the heavy bumper on the backside of Dorothy's camp trailer, which was made out of four-inch channel iron, jammed into the passenger-side door of Bill's new pickup, and Tubby was going so fast that he didn't get stopped until the channel iron bumper had put a slicing gash four inches deep all along the right side of the Ford truck, from the door that was ruined, all the way to the right taillight. When it got to the taillight, Tubby got stopped and the dust settled. The beer was bought and the damage was done.

Tubby showed up at Aso about 4:30 in the afternoon. The cow work had been accomplished in smooth fashion with everyone making a hand. Bill was in a good humor. Of all the men I've ever worked for, he was the master of directing a crew of men and a herd of cows to make sense of each other. The logistics and the placing of the pieces of this puzzle together and making them fit were his life's work. It had been a good day, and the new Ford pickup appeared with short and stocky Tubby Greenhaw at the helm.

It's funny how scenes in situations sometimes unfold; all of us making up the cowboy crew saw it right off, except Bill. An ugly gash, from the right hand door slicing backwards to the rear tailgate, went with two furrows made by the two sides of the four-inch channel iron cutting deep into the sheet metal, plus an extended ditch plowed somewhat slightly less deep in between these two deeper wounds, but all together creating a furrow where pure metal had been stripped of its green paint and thus made shiny and bright, but not joyful, told of a truck and trailer colliding in a cloud of dust after the loading of a case of beer. Nobody said anything, instead we began piling into the back of the pickup for the ride to Spiderweb. It went without saying that Bill would take over behind the steering wheel in spite of the fact that Tubby, who stood five feet three, had been allowed to drive the new pickup, in which the new smell had not even worn off, up to Aso. Bill would drive it home. Home to Spiderweb. None of us wanted to ride in front with Bill, except, of course, Pat Lauderdale, who was always available to be Bill's closest confidant. The rest of us piled in back except for Tubby who just kind of stood there trying to smoke a cigarette. He looked like a rabbit who was about to be shot with a .22 caliber rifle by a 12-year-old boy. Sort of innocent.

And then Bill saw it. The gross violation that the rest of us had noticed five minutes previous. For a moment he just stood and stared like a gunsel county boy who just lost his whole wad at the blackjack table in Vegas. Tubby stood a few feet away with the cigarette held in place by his front teeth, but quivering, going up and down like a butterfly's wings; and his whole body trembled as if he had stuck his finger into a light socket. I have no memory of Bill saying anything, only staring, and then finally he looked at Tubby who said nothing. His trembling said it all.

Bill got in behind the wheel of the now very used Ford pickup with Pat Lauderdale riding shotgun so he could offer condolences or curses or any kind of lamentations needed. Bill put the Ford in gear, and someone riding in the back of the truck reached out and grabbed poor Tubby by the neck as the truck lurched forward. We offered Tubby a perch on top of the spare tire that lay flat on its side in the bed of the pickup. We asked him to recount the application of the truck's horrible and newly acquired blemish.

Poor Tubby tried to talk but was quite agitated. In an attempt to calm himself he would insert a new cigarette into his mouth, pinching it with his front teeth, which were one of his best features, very white, straight and healthy teeth; especially for an old drunken cowboy. He would pinch those cigarettes with his teeth while trying to light them and then as soon as red embers appeared at the cigarette's end, he would start shaking and trembling and would accidentally drop the cigarette out of his mouth. He would immediately produce another cigarette out of a shirt pocket and the whole process would be repeated until finally, before we reached Spiderweb, his pack of Marlboros had been depleted. We never got the whole story. We only knew that a cloud of dust and Hank's Trading Post and a case of beer were involved.

When we arrived at Spiderweb, I went home to the little shack next door to the bunkhouse where Jean Ann and Everett were waiting. Jean Ann and I had moved back down there from Cedar

Ranch a few days previous. This move, like the one going up in the spring had been accomplished in a couple hours in a short afternoon. I was happy to have my own home and family to go to. Because of it I wouldn't have to witness the next stage of the Tubby Greenhaw tragedy.

The next morning broke early, the plan being to eat breakfast and then drive back up the hill to Aso where our horses waited, and then we would proceed to wean 1200 calves. It was going to be a big day. One of the biggest of the year. I walked into the bunkhouse and entered the big bedroom where about eight or ten men had been sleeping in their bed rolls that were scattered about the room and, for the most part, rolled out on the cement floor. Several men were up drinking coffee while several others were somewhere in the process of getting up. The doorway going into the kitchen was open, and Dorothy was busy cooking breakfast. There was a long rectangle table sitting parallel to the wall with enough room between the table and the wall for a long bench.

Tubby was sitting on the bench staring at everyone, looking through the doorway and out into the big room. He was drinking coffee and smoking cigarettes. The cigarettes were held steady in his fingers. I walked through the doorway and into the kitchen passing Tubby on my way to get myself a cup of coffee. I saw Tubby plainly and almost laughed out loud but held myself in check out of respect for his old age. He was black all over as a result of being covered with soot. He was also quite drunk. From viewing his covering of soot, I imagined that sometime in the wee hours of the morning he had tried to make a fire in the cast iron stove that was in the middle of the big bedroom. Perhaps in his drunken state he had actually climbed up and into the stove, hence his covering of soot. He was hilarious with his big white eyes exploding out of his blackened face. I got my coffee and walked into the big bedroom and sat in a chair, and Tubby began speaking. "Look at 'em!" he said as if talking to someone or everyone, "Look at 'em! They think they're cowboys. I'm telling you, there ain't a good cowboy in the whole damn bunch!"

He sat there putting a Marlboro to his lips and taking a long pull on the fire stick as Dorothy stirred gravy on the stove. Tubby's coffee was no doubt spiked with Seagrams VO whiskey that was readily available from a bottle on the big table where he sat. "Look at them! There ain't a good cowboy in the whole bunch of them!" And then the door opened from outside, the same door that I had just walked through when I entered the bunkhouse minutes before. Bill Howell entered the room and Tubby's cigarette fluttered like a hummingbird's wings. He got real quiet. Bill walked past him without speaking. When Bill passed, Tubby stared at him with his big white eyes, and gritted his teeth in defiance. His white eyes and healthy white teeth being about the only thing you could recognize coming out of his covering of black soot. Bill passed by him a second time as he walked into the big room and joined us as we drank coffee and waited for Dorothy to holler chuck.

A short while later, after eating breakfast, we all loaded back up into the used Ford pickup and headed uphill toward Aso. This time we went without Tubby, evidently he and Bill had reached an agreement about Tubby finding employment elsewhere.

A few days later on October 28, 1978, Clay Henry Ashurst was born. He was potbellied with fat smiling cheeks. He was a happy baby and we were happy to have him. Everett now had a brother, a thing that I had never had. I was going to make sure they were friends, although I had no clue how I was going to do that.

Jean Ann's mother came down from South Dakota and stayed with us for a few days to help Jean Ann with the added addition to our family. She helped cook, clean and wash cloth diapers. I have no memory of her saying much, she didn't meddle, or question, or give me dirty looks. If she had an opinion, and I'm sure she had plenty, she kept them to herself, or at least unknown to me.

One day while in town doing business, Jean Ann and I went out to eat lunch at a small café. Clay was a few months old and Everett a little over a year and maybe cutting teeth or maybe had a diaper rash or something, but he was in a bad mood and fussing a lot. People were looking at us, or so I thought. Perhaps, I was just self-conscious, but I remember not enjoying myself.

The café was crowded, and I hurried to finish eating so I could get out of there. I picked up our huge diaper bag, Jean Ann got Clay, and I got Everett, with him still making a lot of unhappy noise. I stood up with the diaper bag on one arm and a screaming kid on the other. Because of the lack of space, the diaper bag knocked over a chair making a racket. People were looking at me. Then I turned and the large diaper bag hit a stack of metal trays. They started cascading toward the floor banging loudly against each other. This caused a chain reaction of one pile of utensils or furniture or something to crash into another, sending a series of percussions that were as loud as a train wreck. I broke and ran for the door with the sound of clanging metal pursuing me. Jean Ann even ran, and we jumped in our pickup which was directly outside and in full view of everyone in the café who watched through a large plate-glass window. Bonnie and Clyde never left the scene of a robbery in a bigger hurry. One of the boys, or possibly both by now, wailed. I almost wrecked the pickup, coming close to running into people, as we exited town in a hurry. I was stressed for several days.

In December Pat Lauderdale, Paul Gonzales and I drove down to Camp Verde and I ordered a new saddle from Scott Dieringer. I had taken my old Bob McRae saddle down to have it relined, and when Scott tore it apart to work on it he found that the tree was broken in three places. A new Dieringer saddle cost about $400, and I was making $450 a month. We drove down to Camp Verde in Paul's new Ford pickup. It was a cowboy rig; a short wheel base with a six cylinder engine. It cost him about $5000 brand-new.

Jean Ann and I went into the winter of 1978 and 1979 in good shape; we had a small savings account in spite of the fact that she gave $25 a month to the Flagstaff Assembly of God church. I was taking care of about 1200 calves in 89 Pasture, and it was so abnormally warm I only had to chop ice in the steel water troughs one or two days the whole winter.

In spite of all the good things happening, and our health and prosperity, I became increasingly restless. I began to listen to the murmurings of a certain cowboy who also worked at the ranch who would make derogatory remarks about the outfit in ever so subtle ways. He would tell me that Bill wasn't being fair with me because I was really his right-hand man, and I should be making more money. He would make sly remarks about all the outfits out there who would like to hire me. He would whisper in my ear.

Bill went off somewhere, I believe Nebraska, and bought 50 new bulls that were about 20 months old. For a few days we kept them up close around the corrals and would feed them a little hay in the big waterlot that was north of the shipping corrals and leading out into the pasture going north. The bulls were able to go out in the horse pasture and graze but would come into the waterlot to eat the hay we would give them.

One day around noon, Paul Gonzales came back from town with a case of Coors beer. We drank several, and then, when we got to feeling real good, we decided it would be a good idea to rope the new bulls that were nibbling on alfalfa right there close to the corrals and barn. Paul and I saddled a couple good horses and proceeded to head and heel the bulls. We knew if Bill drove up and caught us he would fire us both. I have no idea where he was, and I guess we were lucky that he didn't catch us. Luckily we didn't break any of the bulls' legs.

Babbitts had always been a roping outfit. Bill loved to rope. Everyone loved to rope. But roping new expensive herd bulls wasn't something a good company man would have done. I decided it was time to move on to greener pastures. I drove up to Bill's house and told him that I was leaving. Pat Lauderdale was there in the house with him, and I knew there was some whispering going on. When I told Bill I was going to quit, I actually think he was shocked. He looked at me and said, "Damn, Ed, you surprised me!"

I didn't have a job lined up or any idea where I was going to get one. I had been at Babbitts five years, but now I was moving on. It was an old family tradition.

Chapter Three

Jean Ann and I left our home and our job at Babbitts, and after a few days of job hunting, I landed a job at Red River feedlot about 17 miles west of Casa Grande, Arizona. Arizona was blessed with a good spring following a wetter than normal winter, and there were thousands of cattle turned out all over the lower desert regions of southwestern Arizona. The cow market was also very good, hitting all-time high prices after seven or eight years of depression. Fat cattle hit 80 cents a pound for steers ready for slaughter, and 350 pound calves were bringing $1.35 per pound. Red River feedyard was trucking in hundreds of them from as far away as Florida. There were around 115,000 head of cattle on feed at Red River.

A great deal of the time I was at Red River I looked after several thousand steers that were turned out on a piece of desert near the big feedyard. When riding through these cattle we roped and doctored everything that looked sick. If we couldn't find any sick ones we roped steers that were healthy. I worked most days with my friend Jim Dolan, and we got pretty practiced up on our roping skills.

This was before the days of live virus vaccines and much of the acquired knowledge that cattlemen now have about sickness in cattle. A veterinary was constantly advising the feedlot about what medicine to administer to the sick cattle. Sometimes we gave, on a vet's advice, up to four kinds of medicine at once. We gave lots of Penicillin and Terramycin. We did things that cow people now know doesn't work. We created lots of chronics and had a horrible death loss, but we did what the vets told us to do. It was ugly except for the fact that we roped really well, and our horses worked great. I had worked in several feedlots before this and I hated them. I didn't like Red River any better.

There are three things I remember about our time at Red River. The first was an encounter with a little girl on a pony. On a Saturday we went north a few miles to a feedlot that had a big roping arena and they were hosting an exclusive team roping event for feedlot cowboys. It was a good roping and well attended. Jean Ann had pulled up to the arena fence and was sitting there watching the roping while sitting with Everett and Clay, all of them seated in the cab of the pickup. I rode over to them and was visiting with Jean Ann while sitting on my rope horse. Everett was a year and a half, and Clay was about four months. Suddenly a chubby little girl, about eight years old, rode up to my side and stared up at me as I was still mounted on my bay horse. She was riding a small spotted Shetland pony bareback, and her feet were close to dragging the ground. The girl, who I did not know, spoke to me saying, "Are you and your wife going to the dance tonight?" There was going to be a dance held in honor of feedlot cowboys down at the Francisco Grande Resort 10 miles east of the Red River feedlot. I looked at the little girl and answered her, "No, probably not, we have these little babies, and we will most likely stay home and take care of them."

"Well!" she said. "Follow me, and I'll show you something." And she took off in a trot and rode over to some crude pens that had been constructed out of wooden pallets laid against steel posts and tied with baling twine. She sat on her pony and stared over the makeshift pallet fence looking at something.

I rode up and looked down also, and there inside that pen was a nanny milk goat whose kids had been separated from her and were locked away in an adjacent pen. The nanny goat had the biggest udder I had ever seen on a goat. It was swelled up, a result of not being sucked for a few hours, and her teats drug the ground and were squirting milk. The little girl spoke, looking up at me as she did so, "Now, looky here," she said, "when it's time for that dance you just bring those kids of yours over here and stick 'em on that goat, and you and your wife can go to that dance!" With that problem solved the little girl took off in a trot, her behind bouncing on the bareback of the Shetland as she swatted him with a tamarack switch. No doubt she was looking for another problem to solve.

There was a certain cowboy at the feedyard that took a dislike to me, probably due to the fact that I didn't think as highly of him as he thought I should have. We never really had any words of conflict, but rather just an exchange or two of dirty looks.

Each day we would quit at noon and take 45 minutes off for lunch. On this certain day Jim Dolan and I rode up to the horse corral and saddle house and unsaddled our horses, turned them loose in a stall and proceeded to put our saddles into the saddle house, which had a high step you had to negotiate when stepping inside. This cowboy who disliked me was standing in the door doing nothing but watching us, and when I stepped up to the doorway, he blocked my way as he gave me a Lee Marvin dirty look. Packing my saddle as I went, I had to push him aside to gain entry into the building. He said something and I said something in return, and I walked back outside ignoring his Liberty Valence look as I walked past. Jim and I got into his pickup, and we drove to our houses located in the company trailer park a quarter mile away. I didn't think the whole incident amounted to anything, but in spite of that, Jim Dolan kidded me about it, and we both laughed.

Forty minutes later Jim drove by our company-issued house trailer and picked me up. We drove back to the saddle house, and as we approached we observed the angry man standing in the saddle house doorway. He had strapped a black leather gun belt around his hips, and the belt had several dozen .45 caliber bullets held in little leather loops around the back of the wide western-style gun belt. In the holster, which was tied by a string to his thigh, was a Colt single action pistol. The image was exactly like Liberty Valence as he leaned against the post holding the porch roof up and laughingly staring down and mocking Jimmy Stewart as he stood frightened and shakily holding a pistol a mere 40 feet away. My nemesis had his black cowboy hat pulled down shadowing his shifty eyes, with his left arm stretching across the open doorway and a toothpick sticking out of his mouth, which he rolled around with his tongue. "Look at him, Ed. He's got a gun." Jim laughed as we stepped out of his truck. The man stared down at me from his lofty perch on top of the saddle house step. I walked up and ducked under his arm and retrieved my saddle off of its stand and pushed the man's arm out of my way as I stepped out the door. Jim and I caught a horse apiece and rode away leaving Liberty standing in the doorway gazing at us. The next time we met, he had taken his gun off.

One of the cowboys we worked with at the feedlot went by the name of Andy. He lived about two doors south of us in the company trailer park. Andy was a good hand and well liked but possibly participating in some activities involving non-prescription drugs, or maybe he just made some questionable area-residents angry. I don't know or want to know, but anyway, one night as Andy and his live-in girlfriend were sitting in the house, someone drove by and fired a 12-guage shotgun into the living room window and then sped off into the night in an unseen vehicle. Buckshot whizzed through the trailer house along with broken glass, proving that we lived in a very nice neighborhood.

About this time I became very sick and drove into the little town of Stanfield several miles away to see a doctor. The old physician was probably in his 80s and worked by himself in an ancient wooden-framed building that held a few medical instruments and an antiquated examination table. There were some pill bottles and containers of various medical solutions on several shelves. The old man examined me and asked me several questions and told me to drop my pants. He filled a very large syringe, which looked big enough to hold cow vaccine, with white stuff; I suppose it was penicillin. He pumped me full of the stuff and charged me $25. There was no receptionist, nurse, or paperwork to fill out; just an exchange of medicine and cash. The old man had a good reputation in the area, and it was said that he ministered to many low income people, most of whom were Mexican farm laborers, for little or no pay. He was a little man, wore glasses on his balding head, and said very little. He was just a good old man doing what he could to make the world a better place to live in.

The penicillin helped, but I couldn't kick the disease, whatever it was; at least not completely. I decided that I had Valley Fever so I decided to get out of the Valley. I had always hated the southwest corner of Arizona: Phoenix, Casa Grande, Gila Bend, Yuma. I was happy from Wickenburg on north, and on top of the Mogollon Rim, but I wasn't happy below 1500 feet in elevation. I didn't like Red River feedyard, so we moved on. It was a family tradition.

I landed a job on a rough country outfit in the mountains west of Prescott. We were moving to a remote cow camp known as the Halfway House located about 70 miles, more or less, from town. I had been given a foreman's position. It was my first official job as a person responsible for getting the work done on a ranch. I even had a crew to boss, that being one Mexican cowboy who was in the country illegally, not having any kind of legal visa or papers of any kind. He also did not possess any cowboy experience except several years being employed on that place. In other words, I had inherited him with the job. I had to get along with him, but he did not necessarily have to get along with me. I had no power to fire him because the owner was quite fond of him. I did not have the authority to hire any other cowboy help but was told Raul would suffice. Raul did not speak any English, although he had been a resident of the United States for a number of years. I spoke a little Spanish, a result of working with a few other Mexican cowboys and ranch hands who also were in the U.S. illegally. But my communications with Raul were deficient for sure, sometimes being reduced to something akin to sign language or strange guttural exchanges that would have been hilarious to an innocent bystander that enjoyed observing human comedy; but this didn't happen because we worked alone and therefore had no witnesses.

One of the demands of my job was that Jean Ann would be expected to cook for Raul because there were no kitchen facilities in the bunkhouse where he lived. Jean Ann agreed to fulfill this obligation even though I had been told that she would not be paid any wages to do this. This was going to be the first of several ranch cooking jobs that Jean Ann would eventually handle. She took it on with no complaints, but instead she put her nose to the grindstone and did her best. She did it for me, to help me out, so I could be a success at my first foreman's position.

Several funny things happened concerning Jean Ann's cooking for Raul. The first thing was her Spanish speaking ability. I came into the house one day and she told me with great excitement that she had learned a Spanish word. What was it I asked? "Onyons," she exclaimed with great pride.

"You mean onions?"

"No. Onyons."

"What does it mean?" I asked.

"Onions. Raul asked for onyons, and I figured out that it meant onions."

"Cebolla is the correct word for onion."

Jean Ann looked deflated. "It is? It isn't said onyons?"

Raul was still sitting at the table and could follow the conversation, and Jean Ann knew he could follow the conversation. They were both embarrassed, but I thought it a good laugh.

Another time she cooked up a big meal for the owner, his grandson who was 19 years old, plus Raul and our family. We were late getting the big day of work done, so Jean Ann had plenty of time to worry that there wasn't enough food, especially since we were hours later than noon getting in. The owner got into the house and sat watching Jean Ann finish cooking a big pan of fried potatoes, put them in the last free spot on the counter that didn't have some other food or a dirty dish occupying it, and put a lid on them to keep them warm. The kitchen was very limited in counter space.

Everyone got gathered up, and Jean Ann put all the food in one place. We all filled our plates, ate, and the extra people left. Jean Ann began to clean up the mess, and that's when she found the bowl of fried potatoes still sitting on the counter, under the lid. She cringed, thinking of what the owner, with nothing on his mind but watching her, thought of a foreman's wife who couldn't remember to serve all the food she had cooked. I thought it a good laugh.

I was so happy when Ed got a job on a ranch. I was moving away from the feedlot! We had been given one of the nicest trailers in the company trailer park, and the folks before us had kept the lawn green and thick. But it was getting hot. When I hung the boys' diapers on the clothesline, the first one was dry as soon as the last one got put up. And it was about like living in town, and that was bad.

Ed had caught some bug in the Valley and just lay on the bed groaning for days. He decided the air conditioning made him feel worse, so we blocked the vent from blowing into our bedroom. About the time he felt strong enough to try and go back to work, he found a new job. We got boxes, got them packed, and were out of there in a heartbeat. As we drove north, Ed's health steadily improved until when we arrived at the new ranch he was all better.

It was so much cooler up north of Prescott, Arizona than down in the valley by Stanfield, Arizona, it felt like we had turned the calendar back a few months. The first morning when I got out of bed, it was shiver, shiver until dressed. That part never changed because it was always as cold inside the house as it was outside, and as the months passed and the hot pre-monsoon weather settled in, inside the house was as hot as it was outdoors.

Our only electricity was from a generator, and that could only be run when the well was needed to pump water, so running a fan was not an option. I had to carefully schedule cooking a pot of beans or a roast or bread, because I didn't dare get the house any hotter from cooking in the middle of the day or I couldn't stand it. I learned to give the boys a nice, long play-time in the bathtub in a few inches of cool water before their naps so they would be refreshed enough to go sleep.

A few months into our new job the sewer quit draining. We tried several things to fix it, but to no avail, so Ed set his mind to replace the line. When the dreaded day arrived, he started in by carrying the pipe and tools to our front yard and digging up the line. It got hotter and hotter as I hurried around trying to help and make the job easier. Ed acted like it was worse than walking naked through cacti.

At lunchtime I took the boys in to give them a bite to eat and then realized I had stayed out in that heat way too long. Ed did not want to stop and share our lunch because he wanted to get the horrid job over with. The boys and I ate a light snack, and I put them into the bathtub and sat there on the floor by them. Ed came into the house and called out asking where we were. I told him the boys were in the bathtub, come on in. He knew the boys got a bath at night, and he was never around in the day to know what was going on, so he came into the bathroom looking puzzled. I explained what we were doing. He looked at me and said with eyes and with his mouth that I was a genius.

I might have been a genius, but I had not learned to rise up to a challenge and do whatever needed to be done to cope, and I would not get myself into a cool tub of water or put a cool cloth on my head because I hated cool wetness touching me; must have stemmed from a childhood nightmare or some illogical thing in my subconscious. Anyway, after pinning the boys' diapers on and pulling up the plastic pants, I put them down for a nap in their room with the temperature sitting at around 100 degrees. It was extra hot that day. I lay down on my bed and then kicked off my shoes, then I sat up and removed my socks. I lay for a while sweating and then stood up and took of my jeans. Still sweating, the shirt went.

The front door opened, and I heard Ed call for me to come help him. I eased up into the heat, which was hotter up toward the ceiling, and sauntered out of the bedroom into the main room in my underwear. Ed stared and couldn't speak, so I explained that I was trying to stay cool enough so as to not pass out. He still couldn't talk, so I asked him what it was he needed.

"Ah, . . . I need a drink and didn't want to reach into the cabinet with my dirty hands." He probably needed me to come outside and lay in the sun holding something together, or watching for something to happen or not happen, but had changed his mind about getting any help out of me.

I eased over to the cabinet, got a glass, filled it with water, and stretched my arm out full length to hand it to him because heat waves radiating off of his flushed face and sweat-soaked body were enough to raise the temperature around his him by more degrees than I could take. Ed drank his water and watched me as I weaved my 95-pound frame down the hall to disappear from his sight into the bedroom where I lay down again on the slick bedspread, with no pillow so my neck would stay straight, and therefore I wouldn't feel hotter because of the creases caused from bending it. I hoped the boys would stay asleep until it cooled off enough to move without feeling light-headed. I thought about Ed out there in the heat and knew he was a kind of tough that I didn't even have a word for. The next morning it was, shiver, shiver, until dressed.

It wasn't long before Ed came home saying he had taken a job down the road. We were going to have to live in a trailer house/old bunkhouse until the house at the camp where we were going to live was remodeled. The manager had told him it would be ready very soon. Ed mentioned a few things about the camp and house that worried him a little for me. I wasn't worried because I wasn't suffering right at that moment, so I was sure that I was tough enough to handle anything; and a new view out my window and a new place to take a walk with the boys was always welcome.

Raul was haughty and proud and would occasionally make little snide comments about Jean Ann's cooking. He would constantly remind us that we weren't doing things the way the previous foreman and his wife had done them. I was friends with the previous people who had run the ranch and knew that they were very good people. I liked them and respected them, but I couldn't be expected to be a clone. Jean Ann wasn't a clone either. We weren't clone types. In spite of Raul's haughtiness, Jean Ann cooked for him and kept a good attitude.

The outfit didn't have enough horses, but there were several mules to ride or pack. The country was very rough and a considerable amount of packing of salt and other supplies was needed. I didn't want to be unfair and take the best horses and put them in my string, so I let Raul take his pick of what was available and told him I would ride what he didn't want. He informed me that one of the mules was very cranky and might buck a little, but he knew how to get along with him, and he would be glad to ride him and save me the trouble. That suited me fine because I wasn't a fan of riding mules anyway. One of the horses was a dark brown. His mouth showed him to be five or six years old, and he was good looking. I asked Raul about him, and he said he was a good horse and very gentle but also said he didn't want to ride him; so I figured I would try him out.

Early one morning Raul and I saddled up and prepared to go and move some cows. He caught the outlaw mule, and I caught the brown horse. The horse acted gentle enough and I thought nothing about him. I asked Raul if he was well broke, and he said he was. I asked him what kind of bridle to use on the horse. Raul told me that any kind of curb bit would be fine; so I put a bridle with a curb bit on him and led him out of the barn and stepped on him. I noticed Raul had led the outlaw mule a long way off to get on him, and he sat on the mule watching me from a distance. When I got seated on the brown horse, I nudged him to get some forward motion out of him, and he froze up really tight. I could tell from experience that he was going to buck, and he felt like he was going to fall over backwards. He just had a real funny feel about him. I jumped off and took the stiff bit out of his mouth and replaced it with a snaffle. I thought the brown horse might be less apt to fall over backwards with a snaffle in his mouth. I then stepped back on and tried to nudge him forward, but he was still froze up really tight and swelled up. I sunk my right spur as far into his gut as I could send it, and the brown came unglued, and like I suspected, his front end came really high, but I threw a lot of slack into my reins hoping he wouldn't fall over. When he came back down he went to bucking like a scalded dog.

The bucking brown headed downhill, and I was spurring him as hard as I could. We were headed right toward Raul and the outlaw bay mule who went to wringing his tail and spinning. Raul's eyes got as big as saucers as me and the bucking brown came crashing toward them. The mule was half bucking half spinning and always wringing his tail with Raul holding on to the saddle horn with both of his hands. Raul screamed at me, hollering in Spanish to, "Cuidado! Cuidado! Loco cabron!" I spurred the brown and tried to run him into the bay mule while thinking to myself, quidado yourself you sorry son of a gun for telling me this horse was gentle. The brown horse and I soon passed Raul, and we broke into a lope and took off.

We rode to the west and descended off into Burro Creek, a geographical formation very similar to the Grand Canyon of the Colorado. Raul and his mule stayed way behind for a long while making sure they kept their distance. The mule would allow Raul to do anything he wanted as long as the mule approved. Raul kept a death grip on the saddle horn at all times as the mule crept along walking on eggs. Finally I slowed down enough that Raul kept up, and he rode up alongside of me. We visited for a moment, and I minded my manners; at least for a while. Then when Raul got somewhat relaxed and was riding along close to my right side I stabbed the brown horse with my left spur, out of Raul's sight, and the brown horse fell into a bucking fit and the mule whirled and spun almost sending Raul to the ground, but with both hands on the saddle horn, he screamed, "Cuidado! Cuidado! Loco cabron! Cuidado!"

The brown horse would blow up and have a little bucking fit anytime I spurred him real hard, so I couldn't help but have fun with Raul. I managed to maneuver and get Raul out in front of me on a steep trail descending down the canyon wall. The mule would only travel as fast as he wanted to. Raul was keeping a tight hold on his bridle reins, so it was easy to get right up close, with the brown horse's nose stuck in the outlaw mule's tail as Raul nervously sneaked a look over his shoulder wishing me and the brown wouldn't ride so close. I stabbed the brown horse with a spur rowel, and he went to bucking, which sent the mule into a spin right on the edge of a cliff. "Cuidado! Cuidado, loco carbon!" Oh! I loved it. *Complain about my wife's cooking, big shot. Take this!*

We went through the spring with Raul eating with us three times a day in our little scorpion-infested house with a sewer that was plugged up and needed to be dug up and worked on, and other glorious cowboy activities. Jean Ann did not complain about these little difficulties of life. Actually Halfway House was a very beautiful place to live, situated in a rocky canyon surrounded by black malpai rock outcroppings, cedar and pinion trees, and nearly 5000 feet in elevation. Arizona had a good spring in 1979, and cattle were fat everywhere regardless what ranch you were on. Everett and Clay were healthy and growing, and with Jean Ann's constant care, they were never bit by a scorpion or rattlesnake, or a mean dog, several of which we inherited with the place.

Raul and I gathered the cows and moved them all up and out of the depths of Burro Creek, one of the wildest and steepest pieces of country I ever saw. It was all quite easy because the cattle had always been handled correctly, and they were gentle. We branded the calves and scattered the cattle around the summer pastures, and it all got done without any major negative incidents. Raul reminded me daily that my way of doing things was all different than the man who had run the outfit before me. He also reminded me that he got along with that man very well. They had been the best of friends.

Around the first of June, Raul wanted a few days off and needed me to take him into Prescott, taking him into town on a Thursday and dropping him off at the Lido Bar on Cortez Street. We agreed that I would pick him up at the same place on Sunday afternoon. He promised that he would be ready. Raul had made many friends in the Hispanic community in Prescott. He was a sharp dresser and was well groomed and considered himself to be a casanova as well as a rounder.

On Sunday Jean Ann and I and Everett and Clay left Halfway House about noon and headed down off of the mountain toward Prescott, traveling in an old jeep that belonged to the ranch. The jeep had a ragtop to keep the sun off anyone sitting inside, but the sides were all open. It was a very hot June day without a cloud in the sky. We had a very pleasant drive into town, the kids enjoying the breeze blowing into the jeep, which kept it tolerably cool in spite of the heat of the engine coming through the dashboard. Everett was 21 months old, or thereabouts, and Clay 14 months younger. I pulled up and parked in front of Lido's front door and told Jean Ann I would only be a minute or two.

It was a perfect Sunday to be in a honkytonk. The place was packed out. Everyone was having a good time hollering and laughing and drinking many volumes of ice cold beer to offset the sweltering atmosphere that was building out the front door. There was a beautiful blonde gal tending bar who wore a tragic smile that was rumored to be a result of an unhappy marriage which she was doing her utmost to preserve, though a victim of being harnessed to a villain who didn't appreciate her. All of this was talked about in hushed tones by several admirers who wore cowboy hats and were friends of mine. I turned down several offers of a free cold beer.

I found Raul, who was the center of attention in a circle of his friends who were all laughing and slapping their legs at some tall tale he was regaling, all of which was spoken in Spanish. "Vamos, Raul, me esposa y mis hijos ertau esparaudo afuera." I said, interrupting his gesticulations.

"Oh, Eduardo, necesitas relajarte! Tu esposa no va a ninguna parte dejame comprarte una cervesa." His friends all looked at me in agreement, and they all began calling for more cold beer. "Traele a mi amigo una cervesa fria."

"No, really, Raul, we need to go. Vamos." I said.

"No, solo unos minutos te invite una revesa." Raul said giving his friends a smirk as he said it.

I could tell that this wasn't going to be easy, at least not short of an ugly scene. I stood my ground as best I could without being confrontational. "Vamos, Raul. Mi esposa esperando." (My wife is waiting.)

"Deja que tu esposa espere. No la lastimara le tienes miedo?" Raul was now openly sneering at me. His friends looked at me with disgust as they handed me an ice cold long-necked Budweiser.

I was mad and thought about leaving the smart aleck wetback in the joint and driving back to the ranch without him. I knew that Jean Ann was going to be mad as hell, but I drank my Budweiser and visited with several cowboys I knew who were in the place and gave Raul a few minutes to have a few last laughs, and then I walked back over to his huddle. "Vamos, Raul. We need to go!" I gave him my sternest look.

"Ok, Eduardo. Ok. Un minuto."

I got up and headed toward the door. "Let's go, Raul." I was almost hollering. Raul's friends were making disparaging remarks in Spanish that I couldn't understand but their dirty looks were a sufficient interpretation. Raul got up.

"I'm going to buy some beer," he said in his best English. "Un minute," he continued. He walked to the bar and bought two six-packs of Bud Long Necks that were put into a paper sack so they would be legal to take outside, and then he walked back to his huddle and began making farewell handshakes and finally gathered up several sacks of stuff that he was returning to the ranch with. I headed toward the door, and he sullenly followed, carrying his plunder. We walked outside, and the oppressive afternoon heat hit us like the blast of a furnace.

The jeep was immediately out the door parked diagonally with its nose and front bumper touching the curb. Jean Ann's face was flushed with sweat and her hair was damp. Everett, who is fair-complected, looked bleached out, dehydrated, and about to faint. Clay was naked except for a diaper and was crying while Jean Ann held him, which made them both hotter and more miserable. Raul climbed into the little jump seat in the back with a bowed-up countenance and clutching his plunder, especially his beer. Jean Ann wasn't talking. The situation had gone past the point of conversation.

It was about 3:30 in the afternoon, and as I drove by a bank I saw an electronic digital thermometer on a bank marque that read 105 degrees. We drove in silence down Gurley Street to Miller Valley/Iron Springs Road and then turned onto Williamson Valley Road

Chapter Three

and the trail to Camp Wood. I don't remember any conversation in this leg of the trip. It might have been a 105 outside, but in the open cockpit of the jeep, it was about boiling. Clay cried, Everett looked pale and sick, and Jean Ann was sweating. Raul stared haughtily at all of us, and then when we got to the edge of town, he cracked open a cold beer and started drinking.

We turned onto Camp Wood Road and drove a while and then somewhere about the Cross U Ranch, Everett started crying and Jean Ann turned and asked Raul for a beer, and he opened one and handed it to her. She handed it to Everett. "What are you doing?" I hollered at her and snatched the beer from Everett and threw it out the window."

Jean Ann looked at me, and I heard accusations flying, though I am not sure what, if anything, she said. The drive home was ugly, and I could feel Raul smirking about my wife being the boss.

I became increasingly alarmed about Everett and Clay's condition. They truly did look like they needed a transfusion. I drove faster. When we finally arrived at the ranch, Raul got out of the jeep packing what was left of his beer and walked to his apartment in the barn shaking his head and mumbling in Spanish expressing his disgust of American men who don't have the cajones to keep their women in subjection. I had a few chores to do and went to the barn while Jean Ann took our dehydrated children into the house and started ministering to them.

After a while I walked into the house. I noticed that Jean Ann had all the baby food out of the cupboards and on the countertop. I asked her what was going on. She answered that she was leaving. I told her she couldn't just leave, and she said she could leave and that she was going to leave. She said she couldn't handle the lifestyle that went with drinking, and she was giving up trying.

I don't remember what all I said but I argued with her, and then I pleaded with her. I began to think that she was completely serious. I began to believe her. I got worried. We just walked around each other in silence while Jean Ann fed and bathed the boys and put them to bed. Her calm demeanor convinced me that her mind was made up, and I was about to lose my family.

I began to think about what this meant. I began to visualize my future. I thought about the general course these types of situations usually took: The woman leaves and eventually she remarries, and as a result of that, the kids end up with a new dad. I realized that my children would end up being raised by someone else. I realized that they would end up calling someone else Dad. This thought shook me to my very core. I didn't want my kids calling someone else Dad. I realized that I loved Jean Ann but that eventually she would end up loving some other man. That's the way these types of things always progressed. Some other man would be loving my wife, and my kids would call that man Dad. And I looked at myself. I looked down the road into the future and what did I see? I saw myself and it wasn't pretty. I saw a sullen, mean-spirited drunk whose friends had all grown up and moved on to a better life, and I was the only one left in the Palace Bar staring into a beer bottle and feeling depressed and all alone, knowing that my wife and kids belonged to someone else. It scared the hell out of me.

"Come on, Jean Ann, we can work this out," I pleaded.

"If you would get saved things would be different," she told me.

I knew what she meant when she said if you would just get saved. I knew what those words meant. I knew she wasn't talking about jailhouse repentance. I knew she was talking about a serious direction change. She convinced me that there was only one thing that would change her mind about leaving.

I sat staring into the wall and considered the balancing of the scales. I wondered what kind of man I would be if I traded my wife and two sons for a few fair weather friends who wanted to drink and party. I wasn't an alcoholic but I was a party-aholic. Was a party worth trading for my family? I wrestled earnestly with these thoughts for several hours.

Many people have glorious testimonies of fantastic emotional conversions to Christianity. I do not. I have a very sober, even stark, conversion story. I knew that forgiveness was free for the asking, but I also knew that Jesus said, "Go and sin no more." I knew that I would never be perfect, but I also knew that hypocrites don't make it into heaven. It was a conundrum, and yet in my simpleminded state I understood it completely. I was not uninformed.

About midnight on that Sunday night in early June of 1979, I knelt by Jean Ann and beside the bed and asked Jesus to forgive me of my sins. I said something to insinuate that I would start living for him. There were no bells or whistles or cheering crowds or throngs of angels playing harps. I simply walked through the narrow gate and onto the road less traveled.

Chapter Four

That summer the owner of that ranch showed up and spent several months on the ranch in a house located about 10 miles away from Halfway House, the camp where Jean Ann and I and our two boys lived along with Raul who was still eating with us three times a day and living in a bunkhouse a hundred yards distance. Most days Raul would drive the 10 miles to spend the day entertaining the owner who required a partner to help him tinker with what I felt was unimportant stuff. While that went on, I was left to do all the important stuff with no help from Raul who was busy tinkering. I imagined they were also talking about me and how I did everything different from the previous foreman, the good man I had replaced. There was probably no truth in these imaginations of mine, but I imagined them just the same. I didn't like Raul, who in truth was a very poor excuse for a cowboy. I resented his smart aleck attitude toward Jean Ann's cooking. I imagined him sitting on the boss man's porch eating watermelon while I slaved away doing important stuff, and he and the boss man gossiped about me, who as far as I was concerned was the only good hand on the outfit. In other words, I developed a bad attitude.

The boss of the outfit that neighbored the ranch I was running was the same age as me, and he was punchy. He wore a big black hat creased just right with sweats stains in all the right places. He talked about gathering cows in the correct way and doing things the way real cowboys should do them. There were no gunsel chore boys on that outfit, and he claimed he wouldn't put up with the likes of Raul. He said he wondered how I managed to put up with a smart aleck illegal alien who ate watermelon with the boss while I made a hand.

He told me that he was going to need a man to take care of a camp known as the Bozarth Place way down on Bozarth Mesa about 17 miles south of Strojust where he, the boss lived. He told me that the new owner of the Yolos, the outfit he ran, was planning on rebuilding the house at Bozarth, and it would be a perfect place for Jean Ann and me. In the interim time before the remodel job at Bozarth was done, he said that we could live in a small trailer at Strojust, right next door to the house that he and his family lived in.

And so I quit the job I had. A job running a ranch for a decent man who probably didn't eat much watermelon on his porch with Raul who probably didn't talk about me that much. I quit my first legitimate bossing job on a ranch, the type of job I had always wanted, and went to work for the Yolo Ranch for 30 percent less money and moved Jean Ann into a trailer that was 30-foot long by 8-foot wide with a garden hose stuck through a window to supply water for the kitchen. Some of the plumbing in the bathroom worked most of the time. But a newly remodeled mansion was awaiting at Bozarth Mesa. We drove into Prescott and traded our good Ford pickup in on a Toyota Land Cruiser that was older and worn out, because getting to Bozarth was a long and arduous journey. It was also a very rough and rocky road. It was isolated.

Living in the trailer at Strojust was a trying event. It was very small, it was dirty, the plumbing was minimal at best, and it was only several feet out of the back door of the boss' house. We were in the midst of millions of acres of wide open spaces, but two families were living on top of each other.

The boss had a son who was the same age as Everett, and they would play with their toy trucks out in the dirt and would have great fun. Fun, that is, until Everett would decide that he wanted the playing to turn in a direction that suited him better. Everett and the neighbor kid, whose name was Ty, would have an argument and when the tide turned in Ty's favor, Everett would grab an arm or a leg, and he would bite Ty so hard that he would draw blood. Everett's bites would be worse than a little dog bite, and Ty would go running to his mother crying. Ty outweighed Everett by 10 pounds, which is a considerable amount for two-year-olds, but poor Ty had been lectured by his mother to be nice to the little boy that was coming until he was afraid to stick up for himself. Jean Ann would whip Everett good and proper, but the next time Ty didn't play to suit him, Everett would attack again with his fangs showing, and he would draw blood. But the soon to be remodeled mansion at Bozarth was waiting in the wings and only several weeks from being ready to be moved into.

After working at this new job on the Yolo Ranch for a week or two, the boss and I rode horseback down to the Bozarth Camp. I had taken the job without looking at the house or letting Jean Ann look at it, which would have been more appropriate but, instead, completely relied on the boss' description of the place and the construction taking place there.

We arrived about mid-morning on a late August day. There were signs that someone had been around tearing things up, and there were a few tools laying around and perhaps a wheelbarrow laying on its side and a cast off shovel and crowbar, but there was nobody there working. The house was an old, probably built in the 1930s, clapboard shack about 23 feet by 15 feet with two rooms. The main room was 15 feet square. It had been and was going to remain a kitchen, dining room, living room combination. Off to the west side, there had been a bathroom, but it had been recently removed; and there was evidence that a new one was in the process of being constructed. The bedroom lay out on the east side of the main room and was 8 feet wide and 15 feet long. There was a porch that ran along the north side of the structure, its roof being held up by ancient cedar posts and its floor made of planks that had been nailed down in the '30s. Several of these planks were broken or even nonexistent, so if you had a misstep, your leg would fall through the crack and down to the earth three feet below.

I walked around and looked and, evidently, I became visibly depressed. I was no carpenter or contractor, but it seemed to me that there wasn't going to be anything livable about this place until the new year or the next spring. My new boss could feel my depressed vibes and excitedly began trying to pump up my emotions that were spiraling downward. He took me by the hand and pulled me back inside and told me he'd been thinking. He led me into the bedroom, and these words came out of his mouth, "We can have those guys put a wall there," he pointed toward the seven-foot mark on the long side of the room, and he continued, "we can make this into two big bedrooms." He emphasized the word big. I thought to myself: two big bedrooms, each of them seven and a half feet by eight feet. I almost started to cry.

The scorpion-infested house at Halfway was 1000 times better than this, and I had been making a lot more money when I lived there. My memory of Raul softened. I stared off in the distance at the desert landscape and realized that there would never be any electricity here; and it was a 1000 feet lower in elevation. In the summer the old shack that I was standing in would be a 110 degrees inside. I wondered what kind of lie I was going to have to tell Jean Ann when I got back to the trailer house that had a garden hose stuck through the kitchen window so water could be delivered to the sink.

The fall roundup started in mid-September and we, the cowboy crew, moved to Bozarth Mesa and worked cattle for a week or so. There were four of us on the crew: Harry, the boss; a young man named Bruce Liggett; an old codger named Chuck Harvard; and me. Harry insisted on doing all of the cooking. The outfit had butchered a 15-year-old Charolais bull and had 800 pounds of hamburger in a freezer, and that's what we ate; hamburger that was tough like rubber bands that expanded in your mouth as you chewed them. Harry put lots of garlic salt on and in everything, including the fried eggs that he cooked every morning. Occasionally we had a little bacon in the morning. He fashioned himself a sourdough cook, and he made sourdough bread and pancakes every morning. The sourdough pancakes were the consistency of Elmer's glue and would never get any darker than pale pastel yellow because of certain chemicals in Harry's dough that refused to react to a hot cast iron skillet. There was plenty of Log Cabin syrup and so that's what I ate, Log Cabin syrup poured onto my share of the bacon. I waited until Harry wasn't looking, and I would throw the garlic eggs and pancakes into the trash. Lunch and supper was garlic-laced hamburger and sourdough bread, which was really just thicker pancakes that was baked instead of fried, and some kind of potatoes, usually fried. We did have ketchup, so I ate ketchup and fried potatoes and occasionally I chewed on some Charolais bull and got it down. The crew went through cases of Log Cabin syrup and ketchup.

We moved down to the bottom of Scott's Basin where we camped for about three weeks. Between being camped there and a place called Barney Well, we were gone from home for at least a month. When we drove a herd of cattle into Strojust coming out of Scott's Basin, I had lost 25 pounds and weighed less than at any time since I was a freshman in high school.

While all of this had gone on, Jean Ann had held the fort down with the garden hose dangling though the window above the sink.

Chapter Four

She had whipped Everett's butt until it was blistered and poor old Ty had scabs all over his arms and legs. The house rebuilding wasn't making any visible headway at the Bozarth Camp. I never did meet or see the contractors who were supposed to be finishing the remodel job any day. I got disgusted and went to looking for another job. I figured keeping the garden hose from freezing during the winter months would require more ability than my limited engineering skill could come up with.

I got a foreman's job on a huge ranch in southern California, northwest of Needles and south of Las Vegas, Nevada 80 miles. The ranch was about 600 square miles in size and ranged from 1000 feet in elevation to 5000 feet. The headquarters sat at 4500 feet. The headquarters was in the middle of the ranch and was in a forest of Joshua trees, which are large tree-sized plants belonging to the yucca family. These trees have butcher-knife-looking leaves covering all of their appendages, much like the leaves of its cousin the common yucca. From a distance these leaves look soft and bendable, but if you run into one at a high rate of speed while mounted on a galloping horse, it's similar to crashing into a wall of steak knives. It will sure get your attention.

The ranch was known as the OX Ranch and was owned by an older couple who lived several hundred miles away part of the time and there at the ranch part of the time. They were somewhat eccentric and proved to be moody, being positive about life some of the time and depressed or mad the rest of the time.

The ranch could have easily ran several thousand head of cattle but was only stocked with four or five hundred cows plus a few yearlings, this being the result of the owners' financial difficulties from previous years, which ended in large liquidation of cattle numbers and no possibilities of getting refinanced. In other words, the old folks were broke and possessed very little operating capital and very few cows, at least in comparison to the size of the ranch. On a cow ranch, cows are the geese that lay the golden eggs, and on the OX there weren't many golden eggs being produced. However, there were a good many unbranded cattle running around, and the old folks had no real idea how many cows they owned.

When we arrived at the OX there were three employees already working there; all three of which were described as maintenance men and not cowboys. But even with three maintenance men, the place was falling down around their ears. There were at least a dozen water wells on the outfit, but only half of them pumped water; and there was 30 miles of pipeline, most of which leaked water so bad that no water reached the water troughs. The fences and corrals were falling down.

At one time there had been 20 or more good corrals scattered about the ranch in strategic places where some cowboys could pen cattle to brand calves or whatever else might be needed, but when we showed up there was only three corrals on the whole ranch that would hold a gentle milk cow.

One of the employees was a young man from Barstow by the name of Don. He was about 21 years old. He was a large man with blonde hair that needed trimmed and a goatee that had gone to seed. He would sit for hours sharpening a large bowie knife that he wore in a scabbard hanging from his belt. He didn't appear to have any duties he was expected to perform, except to sharpen that knife.

The second employee I met went by the name of Ken. He was from Needles 45 miles to the southwest of the headquarters. He had a wife and home there, and he drove back and forth to the ranch every day. Ken was somewhat capable, possessing a small amount of mechanical skills and knowledge about water wells, and he could have been an asset to the operation; but he was shifty and had a drinking problem and probably a familiarity with narcotics as well. I considered him to be more knowledgeable than Don, who knew nothing, so I immediately put considerable pressure on Ken to start doing some maintenance, which is what he was supposed to do. He would lie to me on a regular basis, and about half the time he wouldn't show up for work, sometimes for several days at a time.

I leaned on him harder, trying to get some work out of him, and he went AWOL more often; and when he did show up at the ranch he would be badly hungover or high on who knows what. He looked and acted like a Hell's Angel or back-alley criminal. He kind of scared me, and about that time in my life I wasn't afraid of much. After a month or so he quit, and I wasn't sad when it happened.

And so I was left with Don and Dmitri. Don was not capable of doing anything on his own. With some adult supervision he could help dig a post hole or dig up a leaking waterline, but by himself he couldn't do much except eat and sit in an easy chair and sharpen his big bowie knife.

Dmitri claimed to be a mechanic and capable of fixing anything, so I promoted him to be the head waterman.

Chapter Four

The ranch was dependent on the water wells that were scattered all over the 600 square miles of desert. Several of these wells supplied water that ran down the country in one and a quarter inch black plastic pipe that was buried a few inches in the ground. This pipeline was old and problematic, prone to get air locks that would completely stop the flow of water, plus numerous and frequent leaks. Coyotes would dig up and chew in two the plastic pipe to get a drink, and the old lines would pull apart at joints and spill the flowing water out onto the ground. So these pipelines were a constant headache and needed daily attention.

The wells were pumped by wind turning the fans on windmills. Some of the wells were a mere 100 feet deep with the deepest being around 500. All of the equipment—towers, mills, fans, pipe, and cylinders—were old and in bad repair. Everything was constantly leaking, and as a result the ranch was always a day or so away from a crisis.

While all of this was going on, Jean Ann was busy raising Everett and Clay as well as cooking. The ranch owners had always had a cook hired fulltime to cook for the crew, as well as themselves when they were at the ranch. There was a small cookhouse that was the first building you came to when you drove into the headquarters complex. This cookhouse was one big room with a kitchen on one side and a big dining and living room area separated from the kitchen area by a bar or counter. Going through a door to the back of the kitchen was a commissary room. It was all functional, but there were no bathroom facilities in the building, and so if anyone had to answer a call of nature they would have to leave the cookhouse to find relief in some other building off in the distance. Jean Ann and I and our kids lived in a trailer house out the back door of the cookhouse, and so with two little boys, she became well acquainted with the trail to our house leading a small child to the bathroom and bringing the other because he could not be left alone.

The owner, an old gentleman named Ed, liked to sit in the cookhouse and observe all of the goings on: the cook cooking, Don sharpening his bowie knife, or perhaps watching the news on a TV that he made sure was in good working order. He liked to sit and discuss the overall ranch work with me, and we got along quite well even though he was very eccentric. His wife was a blast of nervous energy and talk, who obviously likened herself to Barbara Stanwyck in the role of Victoria Barkley on the TV show Big Valley.

She informed Jean Ann early on that her husband, Ed, was the King of the Cookhouse. He played the role well, as did she.

After being there about a month I was able to hire a good cowboy by the name of Bob Grant and was glad to have him. The ranch had wild and spoiled cattle all over it, with many full-grown, unbranded cattle of every size and sex roaming around. No one had got a good count on the livestock in many years, so I was glad to have a good, energetic cowboy to help me subdue the wild beasts who ran around acting feral and happy.

Everett was now two and a half and wanting to get out and move around, so on a few occasions, Jean Ann let him ride along with Dmitri as he checked wells and pipelines. Sometimes I took him out in a pickup to look at some remote part of the ranch.

One day when Everett was gone and therefore not around to help Clay stay content, Clay got mad while Jean Ann was doing laundry, using on old wringer washing machine that was located in one of several outbuildings scattered around the headquarters. Jean Ann had to go from the cookhouse to the laundry house to the clothesline to our house in a constant motion to keep everything moving forward. Clay got mad at constantly being uprooted and moved to another place or left behind. He was in a stage that when he was angry he would bawl and walk forward with his eyes shut. We called them the blind staggers.

Jean Ann left Clay to follow along, bawling and walking, as she hurried toward the cookhouse to stir the beans or some other thing, knowing she would be right back going the other direction and could pick him up on her way back. Clay's bawling turned into an odd squeak before she got through the door of the cookhouse, so she paused and looked back. Clay had managed to get lined up with a baby Joshua Tree, and he was balanced right on the top of it on his little fat stomach, hands not quite reaching the ground on one side and his feet off of the ground on the other. Jean Ann hurried back and lifted him up into her arms. That cured his blind staggers for good.

During January and February of that year, 1980, Bob Grant and I did a lot of riding and were able to gather a good number of spoiled and unwanted cattle that were shipped to market, which created some income for the ranch owners and made them happy. We also branded a considerable amount of unbranded cattle, both calves and some older cattle. In doing all of this, we were getting a little better idea of how many cattle were on the ranch, which was

Chapter Four

knowledge that no one possessed. All of this was fun for me and Bob, and we got along well. It was also work that I was good at, especially when compared to mechanical work which I was not good at, nor did I enjoy.

This period of good times soon came to an end when March rolled around, and the truth about Dmitri began to reveal itself with the knowledge that the ranch's water system had completely collapsed. On or around the new year there were six or seven of the dozen or so wells on the ranch pumping water, but when the first of March arrived, there wasn't a single well that would successfully get water to the surface. Dmitri had been lying about a lot of things, and, I suppose, I wasn't checking on him enough.

Toward the end of Dmitri's tenure as the ranch's waterman several funny things happened. The first came when the big windmill just out the front door of the cookhouse quit working. There was something broken in the gear box up in the tower where the shaft from the center of the big 12-foot fan transferred the wind's energy into the gears that then produced the up and down motion that pulled the sucker rod up and then pushed it down. Dmitri had been telling me that he had the power to control his mind, which in turn could control his body and enable him to do feats that were supernatural. He had informed me that this mind-over-matter ability of his could enable him to jump off of a 30-foot windmill tower and not suffer injury.

Dmitri and I climbed up onto the tall windmill tower with a few wrenches and other objects that we thought might be useful, one of which was a few feet of chain complete with hooks on its ends. Dmitri had told me that he could fix the broken gear box.

The first thing I did when I got onto the small platform near the top of the tower was to take the length of chain and weave it through several of the spokes of the fan and then secure the remainder of the chain to the top of the tower just below the gear box. We also had set the brake on the fan as well, this being done at ground level. It is impossible to completely stop all movement of a windmill fan with the break and a piece of chain, but you can, if you do it right, stop the rotation of the fan and even keep the whole mechanism from turning too much. A windmill fan is designed to rotate and face the blowing wind, this being accomplished by the tail of the windmill which is equivalent to a ship's rudder. A windmill brake is a mechanism that pulls the tail into and parallel to the fan which keeps the fan from facing directly into the blowing wind. At any rate, I chained the fan up as best I could and was sure it wouldn't rotate 360 degrees, but I also knew it might wiggle a little.

Dmitri assured me that he had no misgivings about the fan or the windmill's height or the problems that might exist inside the gearbox, so I told him fine, since you know so much climb up the side of that fan and reach over and unbolt and remove the steel hood on the gearbox and thus expose the gears that are hidden therein. And so he proceeded to do just that, he climbed up the fan and hooked his legs over each side of the top of it which was considerable higher than where I stood in safety on the platform watching him. While perched, like a monkey, on top of the fan he was probably 40 feet off of the ground. Hooking his legs and ankles through the fan's spokes, he leaned his head and upper body out into open space and began removing the hood off of the gearbox.

Suddenly it happened. It was providential, or at least coincidental. A sudden gust of unexpected wind caught the fan and rotated it about 15 degrees and the fan turned six or eight inches very quickly. The whole tower shook and pieces of mechanism unknown to me made sudden eerie, squeaky, ghostly noises. "AAHH AAAHHH AAAHHH!!" Dmitri cried loudly. "Oh shit, stop it, stop it." He clutched his arms and hands through the fan's spokes. He turned white as a sheet. "Stop it, stop it," he wailed.

"Jump, Dmitri, jump. It won't hurt you." I told him. I burst out laughing.

He dropped a heavy crescent wrench, and I was glad that one of my kids wasn't underneath the tower watching. Dmitri did not know anything about fixing the gearbox on the big windmill, and suddenly I realized why there wasn't a single well on the ranch that I was supposed to be running that would pump water.

I fired Dmitri and told Bob Grant to back a company pickup up to the dilapidated shop building and help the crazy Russian load all of his personal tools and then take him home to his homestead, which sat about five miles east of the little village of Goffs that was surrounded by the ranch. Dmitri's tools for the most part consisted of old, rusty, antiquated things you would find in an old machine shop, most of which I didn't think he knew how to use. There were very few simple tools a man would normally use on a big cow ranch.

Dmitri had been driving a pickup owned by the ranch back and forth from his house and the ranch, which was a distance of about 30 miles so I needed to use that same pickup to get rid of him. I enlisted Bob to do this chore also. When Bob got back to the ranch after several hours he came to me and said, "Man, that guy is crazy. I mean really crazy. We were going down the road, and he looked at me and said, 'Bob, don't tell anybody, but I'm from Mars.'" I got a good laugh out of that, but then I realized that I had been allowing my two and a half year old son to go riding around the ranch with someone that was that crazy.

I set to work repairing windmills and made big Don quit sharpening his Bowie knife long enough to help me. On days when I had to have good help I got Bob Grant to help also.

All of the tools the ranch owned were old and well worn: wrenches, elevators, block and tackle, and steel cables and chains. But there was enough there to get something done. To pull pipe and sucker rod out of a well, we would chain a pulley in the top of the tower and run the cable through another pulley at the base of the tower and hook the end of the cable to the front bumper of an old Dodge pickup. The pickup had been a four-wheel drive, but someone before me had taken the drive line off of the front axle so it was only a two-wheel drive. That worked fine on the shallower wells, but on several of the deeper wells, we barely had enough traction to pull the pipe out. This was especially true when we had a check hung in a cylinder and had to pull the sucker rod at the same time along with the water column full of water. Somehow we made it work.

For the most part the problems were simple, wells that needed re-leathered or a pinhole in the water column that needed to be welded shut. Several of the wells needed new brass-lined cylinders. For the entire month of March, I did nothing but work on wells, and when the first of April came I had nine of them pumping water. Several of them I had to pull a couple times, and one well north of headquarters 10 miles had to be pulled and repaired three times, each time being a different problem that, when fixed, would work for a day or two; and then something else would go wrong. I learned a lot about wells in that 30-day period, and it probably didn't hurt me, but I always hated grease under my fingernails.

With some water pumping and some of the many pipelines running water down the lines to water troughs scattered out over 600 square miles, Bob and I started gathering cattle and branding calves as well as big mavericks that we would find. The people before me had gathered the cattle by loading horses into a stock trailer and driving around the ranch, and when they would come to a watering facility where there was a corral, they would unload their horses and try and pen what cattle were at the water trough. Any cattle that would run off were let go and not pursued, at least to any extent.

So, as a result of this process that was regularly used for several decades, the cattle would start running and scattering when they heard a pickup and the rattling sound of an old worn-out trailer that was approaching where they might be laying around. If you approached a water hole where 50 head of cattle might be, 45 of them would take off like birds being shot at, and only the old, lame, or dumb would be left. This method had been used for so long that the sound of a pickup and trailer would make cattle run even if they were several miles from a corral or water hole. Cattle were used to running toward you and then on by you rather than stopping and peacefully turning around and entering a corral like gentle cattle would normally do.

Bob and I would try and make drives starting way out several miles from a water or corral and get the cattle moving in the correct direction, but even then we had many cattle turn and run by us when they got close to the corral. I figured that it would take a considerable amount of time to retrain the cattle and get the spoiled attitude out of them. But Bob and I worked well together, and we got a lot accomplished in spite of the spoiled, cantankerous cattle and the worn-out facilities.

While we were living at the OX we drove to San Diego one weekend and had ourselves a little vacation. We drove into Oceanside from the east and went straight down to the ocean. Jean Ann had never seen the ocean, and, as long as I live, remembering her reaction will always be a favorite memory. We stood on the beach—me, her, Everett and Clay—and the three of them stared at the Pacific Ocean, and I stared at Jean Ann. I'll never forget the expression on her face. She was 23 years old and I was 28. Everett and Clay were too little to know what was going on.

We went to Sea World and the San Diego zoo where Clay chased the little fuzzy animals around, and we laughed while he did it. That was definitely the best part of the trip for him. We also rode on a tourist boat and took a tour of San Diego harbor. That was a good time and something we would do again.

The old couple we worked for were suspicious and always thinking someone was out to get them. One day they called me over to their house, which was a quarter of a mile away from the cookhouse and our house. They said they had something important to discuss. When I arrived the old man sat me down in a chair and then went to his desk and brought back something to show me. His crazy old wife watched closely and intently as he held out an empty .45 automatic Colt pistol brass cartridge in his fingers. I could tell by his demeanor that he thought that the spent cartridge was something of importance. I did not know what to say. I thought to myself, so what? He explained to me that he had found the empty brass cartridge alongside one of the many roads scattered around the ranch. Who was out there shooting, he asked? Was someone planning on doing someone harm? Could that someone be him or his wife?

I was speechless, and the lack of my interest in the brass cartridge no doubt disappointed them. There were several hundred miles of dirt roads on the ranch, most of which was compiled of BLM public land. Neither I nor they had any control of anyone who might want to drive around target shooting. There were probably several thousand empty cartridges scattered about the ranch. So what? At least that is what came to my mind, but they were sure that there was a conspiracy to assassinate someone, probably them, going on. I laughed at the scenario as soon as I escaped their house and went home.

The OX was very short of horses, and Bob and I had a hard time staying mounted because we were doing a lot of riding; the type of riding that was hard on horses. It was a big ranch with hundreds of square miles of country to try and gather cattle in. The fences were down, which meant that cattle were not contained in any one place, plus the fact that the cattle were extremely spoiled. All of this meant that a cowboy had to whip up and ride hard at times (usually) if you wanted to get anything accomplished. I had broke a five-year-old horse the outfit owned that would buck and was pretty cranky, but he was good looking and a good horse, and I had a couple of other horses to ride; but the truth is, we were barely getting by.

One of the horses I had given Bob to ride was an old sorrel horse that was kid gentle and pretty much a plug, but he made a hand on him. The owner's wife considered herself to be an expert cowgirl and horse woman. She claimed the old sorrel to be her favorite horse, but she never rode. Bob rode the old horse for four or five months without her even knowing what was going on. Then one day Bob and I were working some cattle in the corrals right at the headquarters. Old Ed and Nell, the owners, walked out to the corrals to watch us, and the proverbial poop hit the fan. Nell realized that Bob was mounted on her horse, which she hadn't been on in who knows how long, and she threw a fit. Bob was a good hand and not abusive to a horse in any way, but old Nell got mad. She said a lot of nasty things to Bob, who was riding the horse because I had told him to ride the horse. We were working our butts off gathering and working their cattle who someone else had spoiled and previous employees could not gather, but in spite of this she didn't think Bob should have been riding her pet horse. She did a lot of screaming and cussing and stirring up dust. Finally Bob couldn't take anymore, and he told both Nell and her husband, Ed, to take the OX Ranch and stick it where the sun don't shine; and he unsaddled the old sorrel and quit. Now there was nobody left on the outfit except me and Don, who still hadn't progressed much past the knife-sharpening stage. Bob quit and Ed and Nell stormed off to their house in a huff, and I had no help gathering their wild cattle.

Bob Grant had never been to Arizona, at least to work; and I told him about some of the ranches where he might get a job. One of those was the Double O Ranch south of Seligman. He stopped there and talked to the boss, Mike Landis, about acquiring a job. Mike did not hire Bob but did visit with him for awhile, and Bob told him that he had been working for me at the OX. He also said that I was probably unhappy, which was somewhat true, especially after Nell had just run my only good hand off, and that for no good reason. A few days later a very good cowboy who was working for Mike at the Double O quit. Mike, remembering that Bob had told him that I might be unhappy, drove over to the OX and offered me a job, and I accepted it.

On my birthday, July 2, 1980, we drove out of the OX headquarters with all of our stuff loaded in a rented U-Haul truck and headed out toward Seligman and the Double O Ranch. As we pulled away from the OX headquarters, I noticed Don sitting in the shade outside the cookhouse sharpening his Bowie knife. A few months earlier when I had first laid eyes on Don he had been sitting in an easy chair inside the cookhouse sharpening the same knife. Some things never change.

I had worked for Mike Landis two other times before Jean Ann and I moved from the OX to Seligman and the Double O Ranch. Both of those other periods of employment had been at the

Diamond A Ranch when Mike Landis was the wagon boss, or one of the wagon bosses because they ran two wagon crews in those days. That had been in the fall of 1972 and the fall of 1973.

Mike had been considered a good guy to work for during those Diamond A days, at least among the younger cowboys like myself. As the Diamond A wagon boss, at least when I worked for Mike, he had no real management decisions to make. He ran the roundup crew and gathered the cattle the way he wanted and gave orders to his men to suit himself, but he did not have to worry himself about purchasing bulls, or buying pickups or other equipment, or trying to produce a better calf crop out of the outfit's 17,000 cows. He didn't care how much horse feed cost, or what groceries cost or horse shoes. Those were the Diamond A manager's, Jim Lowrance, problems, not his, and because of that fact we poured the grain to the remuda and ate all the apple pie Rex, the cook, would think about baking. Operating costs and the accountant's spread sheet never entered Mike's cranium when he ran the Diamond A wagon. His only thoughts were to ride hard and fast and make lots of dust and create a pretty shadow while smoking his Bull Durham cigarette with the angle of the sun hitting him just right.

At the Double O Ranch, Mike was now required to make considerable more management decisions, and because of this added responsibility, his sole rule for running the Double O Ranch was to never spend any money. A funny example of this attitude change was Mike's theory of putting iron shoes on horses. At the Diamond A Ranch, Mike never clinched the nails when he shod a horse. Clinching and finishing a shoeing job took time, and he was definitely all about being fast. And if the shoes fell off quicker than they would have if they had been clinched properly, so what? The Diamond A Ranch had lots of money and, therefore, lots of horseshoes. At the Double O Ranch, you darn sure clinched the nails and tried to make the shoes stay on, because the outfit didn't have many new shoes, and you coveted and preserved every horseshoe you could find. Horse feed was the same, at the Diamond A we poured feed out like it was free and endless; at the Double O there was very little horse feed and a great deal of the time none at all.

Mike was a real character and loved to play the part of an old-time cowboy. The fact is he had seen a lot of real cowboy happenings in his 50-some years of existence. I believe he was about 51 or 52 when I worked for him at the Double O Ranch. He would constantly remind me and Charlie Gould, a real good friend who was also working there at the same time, that at the JA Ranch in Texas, when he had worked there in 1949, that none of the horses on that ranch had ever been fed any hay or grain. He would say that when trying to justify the fact that he wouldn't buy any horse feed for the Double O horses. I don't suppose that he ever thought about the fact that every cowboy on the famous JA Ranch had 12 to 20 horses in his string, whereas, at the Double O, we had five at the most and usually one or two of them were crippled due to overwork.

I called around Northern Arizona and got myself some colts to ride at $150 a month for each horse, which was a huge boost to our income. Five of these were going to stay with me through the fall roundup, which meant an even bigger boost to our income. These extra horses also meant that I was pretty well mounted for several months. Charlie Gould had an extra outside horse that he was getting paid to ride also, and without these extra horses, I don't think the outfit could have gotten all of the 7,000-plus steers gathered that fall due to a bad shortage of horses for the crew to ride and work on.

One of the company-owned horses Mike gave me to ride was a little sorrel named Jay Bar. Mike really bragged on him, telling me he was one of the best horses on the outfit. Actually Jay Bar had no cow sense whatsoever, but he was tough and could go hard and fast; attributes that Mike admired in a horse. One evening Jean Ann and the boys came out to the barn and horse corral at the Double O headquarters, which were just a short walk away from where our house was. I caught Jay Bar and put a halter on him. Thinking he was gentle enough to handle it, I picked Everett up and sat him on Jay Bar's back. When the sorrel horse looked back and saw the three-year-old on his back, he threw a fit and launched Everett sideways at a high rate of speed; and he landed 15 feet away and skipped across the ground like a flat rock on calm water. I was so mad at myself I could have bit a 16-penny nail in two. Everett cried but lived through it. It was one of many lessons I learned about being careful with my kids, and I wondered why so much of child raising was learned by trial and error. I wondered if my kids could live though my ineptness.

We went to work gathering steers around the end of September with a crew of five cowboys including myself and a drunken cook named Goat Blair. His cooking fit his name. He was of no account and filthy dirty, but in spite of this, Mike Landis liked him. Mike didn't think cowboys or horses needed good feed. After about a month, Charlie Gould and I mutinied one afternoon when Mike was gone for an hour or two. We scraped Goat up off of the ground, him being too drunk to stand or cook, and we loaded him into the bed of a pickup and hauled him into Seligman and threw him out in a pile on the sidewalk in front of the Black Cat Saloon. When Mike got back to camp and found out what Charlie and I had done, he got real mad and started to say something, but Charlie and I told him that he could do without Goat or do without us, whichever he wanted. He decided to get happy.

That first night Goat was gone, Charlie cooked up some beef and potatoes, and I make two cherry pies. The crew hadn't eaten that good in a month.

Mike Landis' wife was a Japanese lady that he had met on Okinawa when he was serving in the U.S. military, and they had married. When he was discharged and came back stateside, he brought his Japanese bride with him. They had two daughters that were good girls. Mike's wife was named Haru, and she was a very nice person and kind and considerate of everyone. Mike had, for whatever reason, become tired of Haru and didn't want her company.

After we had been working for five or six weeks we moved our camp to a waterhole and corral called Daggs. Daggs was five miles up Chino Wash from the Double O headquarters where Mike and Haru and Jean Ann and I lived. The road between Daggs and the headquarters was all on flat ground and possessed very few rocks and could be driven in ten minutes with no trouble and six minutes if you were in a hurry. But Mike didn't like Haru anymore, so he didn't want to go home at night. That meant that I couldn't go home at night either. It wasn't Western to go home at night, at least that is what Mike Landis preached. Good cowboys had no use for the comforts of home, wife or family, or so he said. Charlie Gould lived in a camp many miles from the headquarters and Daggs, and he wasn't supposed to go home either. For the same reason, it wasn't Western!

Mike traded for a counterfeit bucking horse from a stock contractor by the name of Wayne Tallent. I knew Wayne very well and have a considerable amount of memories of my dealings with him, but that will be for a different book. This horse acquired by Mike from Wayne supposedly had quit bucking at the rodeos but was thought to be a good ranch horse. One day I loaded him into a gooseneck and hauled him to Pica on the Double O side of the railroad and went to prowling through several thousand yearlings in a hundred-section pasture. This was about 40 miles from the headquarters by the road. I found a Hereford steer weighing about 650 pounds with really bad pinkeye in his left eye, and I ran him down a rocky ridge among lots of cedar trees and roped him around the neck while traveling at a high rate of speed. The steer hit the sorrel horse hard, and my horse took to me and bucked me off really hard. I hit the ground among lots of malpai boulders, and the fall really addled me. I was never out cold, or at least I didn't recollect waking up, but I was pretty fuzzy. For awhile I couldn't really figure out what was happening. I just knew my horse had run off in the trees and I was afoot. I walked back several miles and found the pickup and trailer and drove back to the ranch and told Mike that I needed help finding my horse. Had I had my wits about me, I could probably have tracked the horse up and found him myself, but my thinking wasn't as good as it should have been, at least for an hour or two. Having to go back over there and find my horse for me really made Mike mad, and he loved to tell people how I had been bucked off which resulted in me needing help.

Another time I got bucked off, I managed to escape Mike's attention and smart remarks. I had several colts that I rode for Cherry Blair who ranched north of Williams. They were all five-year-old geldings. One of them was a little pigeon-toed brown horse who would buck a little. He was kind of an owl-headed thing, but I liked him. The other colts I rode for her were all good colts. So one morning I got on this pigeon-toed brown and left the barn at the headquarters and rode west and traveled right in front of the house where Mike and Haru lived. It was late in the fall, and we had nothing to do but gather up what little remnant was left on the ranch. As I rode past Mike's house, I saw him sitting at a desk in his office, and he was doing paperwork. Mike was not very good at paperwork, and he had his head down concentrating on some serious calculations. He did not notice me riding past.

I rode on a quarter of a mile and came to a wire gate in a fence corner and got off the bronc, opened the gate, and tried to lead the owl-headed bronc through the gate, but he would not lead. I kicked all of tumbleweeds out of the gate, but the bronc refused to pass through the gate. I thought to myself, "I'll fix you!" I stepped on old owl head and spurred him really hard with the intention of making him run though the gate to the other side. Instead the owl head bucked me off and threw me through the gate and then he stampeded back to the barn dragging my bridle reins. I hated to think about Mike Landis seeing the bronc running past his house without me in the saddle. Shortly after the horse ran past, I had to walk past Mike's office, and he was still staring down at his calculations and never noticed me. In five minutes I rode the old owl head back, going west, and remained unnoticed a third time. Mike never said anything to me about any of this, so I know he never saw me any of the three passages.

Right before Christmas, Mike and I were working together gathering remnant, and we were only short a dozen or so steers. We hauled our horses to the west end of the ranch and unloaded them at a water on a pipeline south of Black Tub several miles. After unloading we rode west and through the gate onto the neighbors right at the foot of Cross Mountain on the north side. We rode around for awhile and found two of our steers, both of them being horned Mexicans, weighing 550 or 600 pounds. They were both very gentle.

Mike loved to tell me wild West stories about riding bucking horses and roping wild cattle and other Western pastimes. I said to him, "Mike, have you ever necked two wild steers together and drove them out of a piece of country tied together?"

"Oh, yeah. Lots of times. By golly - sure works good, too. Yes, sir. That's a good way to do it," he said.

"Man, I would sure like to see that! I'd like to know how to do that correctly."

"Well, by golly, one of these days I'll have to show you. It works real good."

"Do you think we could rope these two steers, and you could show me right now?" I asked.

"Well . . . sure . . . I reckon so."

These two Mexican steers were walking toward the gate like a couple of old milk cows, but I roped one, and Mike heeled it, and I tied it down; and then I roped the other and drug it up to the first one. Mike proceeded to tie the two together with a couple of piggin strings, explaining to me all the while about how to accomplish this correctly.

When tied together the way Mike wanted, we let the two steers loose. The gate going out onto the Double O was only a short distance away. Where we started out with the tied steers there were no trees, but on the Double O side there was a cedar thicket of biblical proportions. There was a good two-track road down through this thicket that the two gentle steers would have walked down to the truck not taking more than an hour to do so, but instead, because of their being necked together, they wandered through the thicket and forked no less than a thousand trees. They were down as much as they were up and getting down to the truck and corral took four or five hours instead of 50 minutes. "By golly, this usually works real good," Mike kept telling me.

When the steers on the Double O had all been gathered there wasn't much to do. I had turned back all of the outside horses I had ridden for other people and had been paid for them, which was a good little bonus; but our future on the Double O Ranch was uncertain. The year of 1980 had been a disastrously dry summer throughout Arizona, and the cow country around Seligman looked like the Dust Bowl in Oklahoma in the 1930s. The high prices cattlemen had enjoyed the previous year were dropping. All of these factors were ingredients in the decision by the people we all worked for at the Double O to liquidate their operation. Therefore their need to keep me and my friend Charlie Gould on their payroll ended. And so for a Christmas bonus, we received the news that on January 1, 1981 we would be unemployed. Mike Landis would continue to get paid so someone would be on the place until someone new leased it and decided to stock it with cattle, but no one knew when that would be or who would do it.

I heard through the grapevine that the Diamond A Ranch needed to hire someone to run a bull dozer to clean dirt tanks on the ranch. I had a little experience operating a bulldozer, so I drove into Seligman and inquired about the job but was not hired.

Word gets around, and sometime in early January, Bill Howell sent word for me to call him. We were still living at the Double O headquarters, but were not on anyone's payroll. I called Bill, and he told me that he needed a wagon cook and wondered if Jean Ann would be interested in the job. If Jean Ann wanted the cooking job, he would be willing to hire me also, but it was plain that he needed a cook and didn't really need a cowboy.

We had been gone from Babbitts for 23 months, and I had worked at five different jobs, having to pack up and move each time. Two of the jobs had been good-paying foreman's jobs that I had quit. I was following my father's footsteps, and moving and quitting was a family tradition. Lucky for me, I had a wife who could cook. Lucky for me, Everett and Clay had a mother who wasn't lazy.

Chapter Five

We moved back to Spiderweb on the Babbitt Ranch in mid-January 1981. There was much to be thankful for: the boss bought as many horseshoes and nails as a man could ever want, there was no shortage of horses to ride and feed to feed them, the outfit never, never ran out of beef to eat or flour to make biscuits and strawberry jam to top them off with, the outfit had good cattle, and fed and took care of them. There was never anything fancy or extravagant about the Babbitt Ranch. There were no pumping oil wells to finance elaborate entry ways into the headquarters, but it was well run. They had always had a maintenance program, and because of that the fences were up and the roofs over the houses didn't leak. I was glad enough to be back to question myself as to why I had ever left. I wouldn't have to stay in a teepee at Daggs while my wife and kids were five short miles away just because some old cowboy hated his wife and dreamed about the JA wagon in the Palo Duro Canyon in 1949. I didn't mind camping out in the middle of nowhere when it was necessary, but the Daggs experience had ruined my concept of the old West. And now we could all eat Jean Ann's cooking. I still had nightmares of drunken Goat Blair.

Bill Howell wasn't big on making speeches but he did mention to Jean Ann that the grocery bill had sort of got out of hand, and he would appreciate it if she would cook good basic cowboy groceries: steak, roast beef, potatoes, beans, homemade bread, and maybe a dessert once a day, homemade things like cinnamon rolls or apple pie. He liked the idea of having a lot of steak and biscuits and gravy for breakfast. He never insinuated that he was going to hold her to a budget or that she had to scrimp and save or worry, just keep it simple and homemade. Cowboy stuff. No one ever went hungry when Jean Ann cooked, and it was always good, good enough to brag about; but she still managed to cut the roundup grocery bill in half. She could really cook. It was a family tradition.

In those days Babbitts always went to working cows around the tenth of February, moving cows out of the winter country east of Highway 89 and heading them west toward the summer range, and so that's when Jean Ann's cooking duties began around February 10. Several days before that, several long-term Babbitt cowboys cornered me in the bunkhouse and informed me that my wife had better cook good and right down to the crew's specifications or they would run both Jean Ann and me off. They had bad-mouthed the previous cook and stabbed her and her husband in the back so much they had quit and moved on to another job. I went home and told Jean Ann that I wanted her to cook to suit herself and to never pay any attention to anything some self-important complainer said. If they went too far in their self-importantance, I planned on whipping their _ _ _. There was never any complaining. Jean Ann cooked good and conducted herself with class. She didn't gossip, didn't get involved in any management discussions, and she acted like a lady—a hard-working lady; and everyone got plenty of good homemade groceries, and it was clean and worthy of compliments. There were lots of them.

Jean Ann would have to get up early and go to the bunkhouse to start breakfast. Sometimes I went a little earlier to start a fire in the wood cookstove and start the coffee, but usually we would go together or her first. We would carry Everett and Clay wrapped up in blankets and lay them on the floor of the commissary room in the Spiderweb bunkhouse where they would sleep through the hurry of getting breakfast ready as well as the eating of it. Sometimes they would wake up and watch big-eyed as the cowboys filled their plates with biscuits and gravy. Homemade biscuits and gravy, the gravy being made in the frying pan after Jean Ann had cooked steak. She could make the best biscuits and gravy ever made. Everett was three years old and Clay was two when Jean Ann started cooking for the Babbitt roundup crew. They grew up around men like Bill Howell, Pat Lauderdale, Tracy Dent, and their uncle Harvey Howell as well as others who drifted in and out of the outfit. Everett especially loved Tracy Dent and called him ol' Tracy.

One fall Bill hired a buckaroo named Kent Craven, and Kent owned a cranky Queensland Heeler dog. Bill Howell was never a dog lover, and Kent was the only man I ever saw Bill allow to have a dog with the wagon. Evidently he liked Kent, and he was a fine fellow; but Bill would often quote Raymond Holt who was famous for saying, "Yes, yes, this would be a fine outfit if everyone had a stud horse and a dog!"

Kent would always keep his mean dog tied up, and one day he was tied up in front of the Spiderweb bunkhouse; and Everett came innocently walking by, and Kent's dog attacked and bit him really hard on his butt. The bite drew blood and scared Everett to death and embarrassed Kent. Somehow we all remained friends, and Everett learned to walk way around mean dogs.

Kent was really good at telling stories, and he kept us entertained with funny cowboy tales of things he had seen. One night in the Redlands bunkhouse, I saw Bill Howell on his hands and knees pounding the cement floor and laughing so hard he was in tears listening to one of Kent's tall tales. He was really funny, which is a valuable thing to have around a crew of men who work hard and get little time off.

One winter, when Everett and Clay were four and five years old, Jean Ann and I decided to take a day trip to Grand Falls on the Little Colorado River. This waterfall is actually only a few miles, perhaps 20, southeast of Spiderweb, where we were living. It had been a wet winter and the Little Colorado was running water, which is not always the case.

Grand Falls is not exactly a tourist attraction, although they are taller, at 185 feet in height, than Niagara Falls. I imagine the fact that the Little Colorado, at least at this spot, is dry and nothing more than a huge sand wash a great deal of the time has plenty to do with its obscurity and lack of fame. But on the day of our adventure the river was running big, being fed with rain and snowmelt from as far away as the White Mountains near Springerville, Arizona, 150 miles away. Claiming to be 185 feet in height is somewhat of a misleading description because in reality Grand Falls are a series of falls that drop 185 from highest to lowest over a stretch of 40 or 50 yards; whereas Niagara Falls and others are one sheer drop off, which creates a more spectacular scene; but Grand Falls when the river is running big and wide is still a fine thing to behold.

Chapter Five

To get to Grand Falls you travel out Highway 15, or the Leupp Road as it is known locally; and then when nearing the river, you turn and go north on a very primitive dirt road and travel on this for eight or ten miles. There are few road signs and a great deal of desolate country to get lost in if a traveler is a novice about getting around in remote places. The country along the river has a great many basalt, or malpai, outcroppings rising out of miles and miles of desolate sand dunes. This is the entry into what is known as the Painted Desert where famous photographers take pictures of Navajo Indians herding sheep in country devoid of anything except sand. Perhaps Navajo sheep know how to eat sand and survive. I'm not sure about that. At times, especially about sundown, the distant cliffs and sand dunes of the Painted Desert create a beautiful panorama that runs for countless miles off to the east of Grand Falls in Navajo land.

After turning off the lonely highway that goes northeast toward the Navajo outpost of Leupp, we bumped along on the sandy road and finally reached the Grand Falls. There were no shops where hucksters sold trinkets to tourists, no paved parking lots and traffic jams, and no park rangers wearing badges; but, instead, millions of acre feet of brown muddy water tumbling down off a series of cliffs that make up the drop of 185 feet in elevation. There was brown mist rising up from the splashing of tons of water being poured over the rocks where it landed, and a loud roar the flood created as it fell to its resting place. The best part about the whole thing was the fact that we were all alone, at a legitimate natural wonder that had never been commercialized.

We parked our pickup and walked to the edge of the first drop off, and we all stood in awe of the beauty of brown water falling downward into the abyss. As we walked along in the direction the water was falling, the riverbank we walked on remained at the same elevation, whereas the river itself dropped off of the face of the earth; and so soon we were on top of a cliff looking off at a great height at the tumult below.

Clay looked like he usually did at that age. He had a grin on his face, and his shirttail hanging out and his little pot belly leading any way he might be headed. He walked close to the edge of the cliff and hung his head forward, straight up from a sheer expanse of open air below him. I watched him warily, and perhaps I even said something to the effect that he should be careful. He was never careful. We continued walking and enjoying the spectacle.

Suddenly I noticed Everett was missing, and I looked around and then saw him. He was sitting in the cab of the pickup, and he had closed the door and rolled up the window. Jean Ann and I walked back to the truck and asked him. "Everett, what are you doing? Get out and look at the river. Are you all right?"

"No," he announced.

"What's wrong? Get out and look at the falls. What in the world is wrong with you? Are you sick?"

"I'm not going to look at the falls," he said.

"What in the world is wrong with you?" I demanded.

"I'm not going to watch my little brother kill himself because you are such sorry parents you won't take care of him!" he answered.

"What are you talking about?" Jean Ann asked.

"You guys are letting him walk too close to the edge, and he's going to fall off of there and kill himself!"

Clay, with his little pot belly and shirt tail hanging out and grin on his face, was on down the river bank staring off of a cliff; and the other one was in the car bowed up and mad because we were bad parents. "Boy," I thought to myself, "this parenting is hard!"

On Christmas of 1981, we were living in the little house that had originally been the first house Jean Ann and I had lived in, the one I referred to as a shack. It was a shack. In 1979 they tore that shack down and moved two rooms of it over 50 yards toward the horse corral. Then they built a new house on the spot where the shack had stood.

We were living in that little two-room house close to the horse corral, and at Christmas time it got really cold. I had bought something for Everett and Clay, but I don't have any memory what those things were.

Ben Fancher was working at Babbitts and living in the Spiderweb bunkhouse. He had a family of his own who were living in Kingman, and Ben would go and see them when he could, but he was employed at Babbitts. One day several days before Christmas, Ben went into town and showed back up at Spiderweb late in the afternoon. He came over to the house and brought some presents he had bought for Everett and Clay. He had bought them two toy rifles. The guns were long, almost full scale, and they had wooden stocks including wooden forearms that ran way out toward the ends of the barrels. The barrels and actions were made of metal, and they had hammers that could be pulled back and cocked. They were the neatest toy guns I ever saw for a kid. They made what I had got for the kids pale in comparison. That didn't bother me because the boys loved the guns, and Ben loved giving the guns to them. I think it made his whole Christmas. It is a happy memory.

When Everett was about five years old, I had a horse in my string named Strawberry. The horse was a beautiful strawberry roan; and as fast, quick and athletic a horse as I ever rode. He was also an outlaw who was very treacherous about a rope and anything dragging from the saddle horn. He would whirl away and stampede out of control if connected to something that was dragging by a rope. He was fast and crazy. I finally roped some big fat dry cows tied hard-and-fast to the horn, which anchored him to something too heavy to drag, and he got usable; but he was not a kid horse and never became gentle.

One morning I walked out to the Spiderweb saddle house with Everett following me wanting to be a cowboy. He had on his little boots and cowboy hat. I got a bridle and walked out to the corral and caught Strawberry and led him up to the saddle house door and handed Everett the reins and said, "Here, hold my horse for a minute." I then stepped inside to get a curry comb or my saddle or something. I only had my back turned for a minute. When I got back to the door where Everett stood, he had taken a bridle rein and threaded it through the belt loops on his blue jeans. He was about to feed the end of the bridle rein through the last loop while Strawberry stood there calmly. I almost died of a heart attack, but instead of falling over dead with panic, I calmly reached down and unthreaded the bridle rein and managed to get Everett disconnected from the worst stampeding outlaw I ever was around.

I don't remember saying anything, but instead I sat down and held Everett and breathed deeply for awhile.

    The winter of 1977-1978 found Ed and me living at Babbitt's headquarters. Pat and Sherry Lauderdale were living at Savage Well Camp, which is beyond headquarters to the east about 10 miles. They frequently stopped in to visit, mostly because their little girl, Johnnie, always begged to go see the baby, meaning Everett who was nine months her junior.

    One evening Pat and Johnnie stopped by just after I had gotten home from town, and shopping bags lay heaped inside the front door. Pat, Ed and I visited as I put away groceries, and Johnnie roamed around looking for ways to prove how much bigger she was than the baby. She swaggered by a box of laundry soap and jerked up on the plastic handle. It didn't budge. Pat and I grinned at each other as she sauntered on trying to show that she hadn't wanted to move the box at all. She was just checking the handle. When she came back by, she had one of Ed's Romeo slippers in each hand. Pat was on to that. "Well now, Johnnie, those boats of Ed's are sure enough a challenge. Some half grown men can't pack Ed's shoes around." Pat had scored one and he was loving it. Ed, one of Pat's favorite victims, just smiled and bided his time.

    After a while Pat got around to asking Ed for a haircut. The clippers were brought out and an old sheet tucked around Pat and the barbershop was open. The clippers, which worked fine the last time they had been used, wouldn't do anything but hum when pushed against Pat's wiry curls. First Ed tried oiling them and then resorted to turning them up until they roared and clattered enough to deafen me as I passed back and forth to the pantry. Every time Ed would try to cut a tightly wound lock, the roar would wind down to a faltering hum; then he would pull them back and go at it from a different angle. This had been going on for at least five minutes without much, if any, hair falling free. I was becoming immune to it all when a violent flinch and screech from Pat jerked my attention to the back of his head. There was a wide white slash in the salt and pepper mat with a fine pink line down the middle. Pink from fresh blood! Pat recovered himself quickly with a, "D_n, Ed, you're getting a little close, aren't you?" I looked at Ed's face. Remorse, well, no. A slight smile rested on his lips and a twinkle brightened his eyes. Score one for Ed.

We moved away for a couple years and came back for me to have the cooking job on the wagon. Everett and Clay were three and two years old by this time, and did they ever love being around the cowboys, breakfast, lunch and supper, and every chance in between. Pat was still working there, and the boys were just the right age to recognize and appreciate a true personification of the wild-west when they saw one. Pat always had a big feather in his well-broke-in hat, and his spur rowels rang out loud and clear as if they were telling about all the broncs they had spurred and wild cattle they had gathered. He visited with them just like they were full grown cowboys and even took them with him when he had a little-kid-friendly chore to do. But the very best of all was the unique way Pat started his pickup. They watched from a safe distance as he stood in its open door and pushed until he got it rolling. Then he would jump in and just when they were wondering if it would work this time - bang, a loud explosion - whoosh, a ball of black. With a leap and a lurch and a roar, he was off. How could a boy not jump from the sheer excitement of it.

One spring Pat decided to make some chaps. He asked Ed if he could barrow his chap pattern. Ed had been making chaps for five or six years and had gradually made up his own pattern and found a source of perfect chap leather that wasn't too expensive. When Pat didn't choose to buy the kind of leather Ed always recommended and didn't choose to ask Ed to make the chaps for him, it freshened old, bad memories of when Ed was proudly wearing his first pair of self-made chaps, he heard that Tom Reeder had announced in his stuttering way that he would make his own chaps. They might not be too good, but they would be better than those ones that Ashurst had ca ca ca cabbaged out. Ed was feeling stung, but when he walked into the bunkhouse to see how the chap making was going and found Pat on the floor with 28 square feet of leather wrapped around himself trying to make sure he didn't cut out two right legs, the score immediately felt even.

A few years later when the wagon was camped at Cedar Ranch, Bill Howell told Ed the boys could go with the crew to brand some calves. That morning the cowboys had penned the cattle at CO Bar Tank on top of the rim above camp, and then had ridden down for lunch. A couple of men would ride horses back up the rim to do the roping and the rest of the crew would ride up in the branding pickup.

Hee Yaw! The boys had thought it was an ordinary day, but no; it was a 10—a perfect 10. They waited where they were told to wait as the crew hustled around getting things ready. Then Ed put Everett in the back of the pickup as the other cowboys jumped in over the sides. He picked up Clay and started walking toward the front as he explained that he couldn't keep two boys safe on such a rough road, and Clay would have to ride in front.

It couldn't be true. From a ten to a zero in one unexpected blow. So what if Everett was a year older. He was no bigger and one year was of absolutely no significance. It was favoritism, pure and simple. Clay walked to the middle of the seat, leaned back with chin tucked and chest swelled. Everyone felt terrible.

Bill slid into the driver's seat, and Pat slipped in the other side. Pat, reading Clay like a book, said with just the right amount of casual attention, "So, Clay, you're coming with us to brand today." Now Clay would have rather hit someone than speak, but this was Pat, so he loosened up enough to look around. Here he was up front between the boss and the wild man. His expression lightened as he twisted his square frame just enough to look through the back window and declare, "So big shot Ed has to ride in back."

Watching from the cookhouse doorway, I could only wonder why the pickup bucked to a stop, both doors flew open, and both men staggered out laughing until they had to hang on the sides of it to stand up.

In the fall the Babbitt wagon moved to Kendrick Park while the cowboys gathered the cattle off the Wild Bill forest allotment. Out on the flat was an old two-room shack. One room was used as a kitchen and the other for the cowboys to sleep in. An old camp trailer was temporarily pulled in as a place for the cook to sleep. It turned really cold the September of 1981 when we moved up to the park. It felt like winter at an elevation of 8000 feet, but the wood cookstove, with wheels on one end so it could be moved around with the wagon, warmed the shack comfortable.

Our first morning there, Ed and I got up early and left Everett and Clay sleeping in the camper. Ed slipped out often to check on them while I cooked breakfast. After breakfast as it started to get light and the cowboys clanked and banged spurs and doors in and out of the cabin, I increased the trips through the cold wind to make sure the boys didn't wake up in a strange place, alone and scared.

I had just returned from a peek into the camper when Ed came in to get some extra clothes because of the cold weather and to tell me that Ol' Pete Michabach had showed up to help gather cattle. He said Pete had been standing by the horse truck visiting with Bill Howell about the day's work wearing a light jacket, unzipped and flapping in the wind. His pants, as usual, sagged down around his hips in a careless fashion. He occasionally gave them an absentminded tug that didn't seem to do any noticeable good. Probably because of the batwing chaps that were too small, buckled too low, and helping the pull of gravity way too much.

The cowboys were astonished at how little he seemed affected by the cold, but when his pants dropped to his ankles, and he didn't even stoop to pull them up until he had finished making his point, they were impressed. The wind whipping his boxers around his 70 plus year old legs chilled every man present to the bone.

The falling of Pete's pants had long been expected, but its realization started the day with a good laugh. There in the kitchen as Ed's story reached its climax and merriment filled the room, the door opened and in stepped Paul Gonzales. His black hat was pulled low against the wind and his gray coat was buttoned up to the black wild rag wrapped high and tied tight. Even his chaps and boots looked dark and stiff with cold, but in drastic contrast, across his chest was a splash of color—flaxy-haired Everett covered head to toe in yellow jammies.

"Everett was crying!" Paul said in the tone of voice young adults use on a parent who is so callous as to be caught laughing when their child is in need. Now Everett looked so peaceful and secure held tight by one of his favorite cowboys that I might not have taken Paul's admonishment to heart but for the big, remnant tear still resting on Everett's cheek.

With many "thank yous" to Paul and "I'm sorrys" to Everett, I took my son and wrapped him in my coat. I figured we had better go see if all the commotion had awakened Clay. As we dashed across to the camper, the horse trucks rolled by, and Everett watched with his usual longing-to-go-with-them expression. I felt a shiver pass through him. "Mom," he whispered in awe. "Ol' Pete has his window rolled down."

"Well, hello, Clay. How are you doing today?" Bill Howell paused for an answer. Maybe Clay would have answered, given enough time, and maybe he wouldn't have, but his solemn stare prompted Bill to rush on with another question. Possibly he hoped this one would be considered more agreeable. "What'd you have for breakfast?"

Clay stood with rounded stomach gently pushing against the front of his homemade shirt, it's tail, made square by Mom because it was always untucked, hung neatly over his elastic-waisted pants. "Cees wif bue fosting."

Bill's face hardened into serious concentration. He had raised three boys and probably thought he had heard it all, but the pressure of Clay's serious gaze seemed to unnerve him. "Well now, Clay, I don't hear as well as I used to. What was that you said?"

Slightly louder and slower, "Cees wiff bue fosting!" The expression had gone from serious to intolerant. Everett and Clay slept through the cowboy's breakfast and had theirs later. Bill had no way of knowing what Clay had eaten for breakfast, but he seemed determined not to offend this cranky two-year-old about his enunciation. He looked at me, silently pleading for help.

I couldn't resist. I said, "Cees with blue frosting." The look in Bill's eyes was sort of wild and trapped, like he could really identify with Dorothy when she landed in Oz. It prompted me to have mercy. "I make Clay's pancakes in the shape of a C," and when he still didn't seem to get it, I added, "and Everett's in the shape of an E. I found some blue decorator frosting in a tube left by a previous cook, and I let the boys put it on their pancakes instead of syrup."

"I see. Cees with blue frosting. Well, that's just what you said, Clay. I understood you just fine. But, well . . . I didn't know your mother was taking such good care of you." Bill might have been my boss right then, but he was also my brother, and I knew that expression. He felt brilliant. He had reassured Clay of his speaking ability and hadn't even needed to lie.

Clay never moved his lips, only his eyes scoffed, "About time you got it right." And he turned on his heel and marched off. I pretended not to notice his icy attitude and hoped Bill wouldn't take it too much to heart.

Bill Howell, warned me a day early so I could get prepared. "I'm going to have Mike go with you tomorrow when you bring us lunch. The road to North Pasture Tank is overgrown, but I think you can make it alright. We don't need much, and don't bother to bring coffee. We'll get by with water."

Some of the good roads I took lunch over would almost turn steak into hamburger so the next day I packed everything extra tight. The roundup pan (the name given to the dishpan used to collect the dirty dishes) pressed the water jugs into the corners of one chuck box. Bread, with foil crimped tight to keep out the dust, filled another. Steak and potato salad wedged together with no shifting room completed the meal. The chuckbox used to hold the pickles and such had to stay at home, but I slipped the salt and pepper into the box holding the plates, cup and silverware, and we were all set to go.

Mike Lenton, the old, one-eyed waterman who had worked at Babbitts most of his life, climbed in the passenger seat as I lifted Everett, serious and responsible at five, to climb in and stand by Mike. "Hello Clay," Mike said. It was Mike's goal to get the boys to tell him that he had called them by the wrong name. Mike never won the contest.

"Hello Mike," Everett drawled as he spread his feet slightly to brace himself against the back of the seat.

Clay came next, a year younger, stubborn and reckless; he always had to stand by Mom. With Clay in the middle, giving me plenty of room, I lifted my foot up about waist high, got a good grip on the steering wheel, and lunged up and in. I wondered to myself if all the clearance under this ton truck was really necessary? The old green chuckwagon's weak breaks held as we eased off the steep slope by the Cedar Ranch cookhouse and made the sharp left turn in front of my house. We were off and running, right on schedule.

In just a quarter mile the road forked. Mike mumbled, "I think we'll go to the north."

I glanced over at him and thought, "You think? If you don't know, who does?" Oh well, Mike always hinted at things. It was his idea of good manners so I didn't pay it much attention. Mike continued his uneasy ramblings and through them I learned that he had decided the old road was too overgrown and we should try to get there from the opposite direction.

Soon it was too late to go back and still get lunch there on time, but should we—could we—go forward. Disregarding Bill's suggested orders had been a risky decision, and now here we were on a malpai ridge squeezing through cedar trees so closely spaced that the mirrors were pushed up against the truck on both sides. We pressed on with Mike making all the decisions. That was fine with me and fair, too. After all, it was his pride and hind-end that were on the line.

I was staring hopelessly around when Mike's voice took on more confidence than it had possessed all morning "There," he said pointing at a break in the trees. "I think we can get down right there."

"Where? You mean, there, in that little break by that big tree. It looks too narrow and the drop off those rocks too high." Visions of axles and mirrors littering the forest floor filled my head.

"I think it will be all right. The best way is to go straight off—real slow." He got out to push back some large limbs and encourage me over the precipice.

I maneuvered the truck into the desired position. "Sit down," I mumbled to the boys as I resisted the urge to close my eyes at what was ahead. Ignoring me, Everett and Clay continued to look around like little owls. Taking lunch out had never been like this, and they didn't want to miss a thing. I reached over and gave Clay's always available shirttail a tug as the building tension filled my voice. "Sit down!" They sat. I eased over the rocks—front tires over, bump, no crash, no scraping; back tires, bump, no bang or clatter of the muffler coming loose. I stopped, the boys jumped up and Mike climbed in. That major hurdle was conquered.

We wove our way through a few more trees and discovered that we were coming into the drive right along with the cows and cowboys. Mike, making the best of this ridiculous situation, guided me to take my place between a couple of the cowboys and idle along beside the cattle. He became silent, but the frozen half grin told how badly he had hoped that we would slip through unseen, leaving only our tracks to be wondered at. We could see Bill looking back at us. His glare clearly hollered, "What is the chuckwagon doing wandering around in the middle of this drive!"

Oblivious to Mike's embarrassment and the cowboy's wide-eyed surprise, the boys pressed onto my lap, side by side on their knees, putting their heads out my window and looking around in wonder. So this was what cowboys did on drive.

The beef in the pan and the beef on the hoof got to the holdup just fine. The humans were a little mixed-up. Mike was chagrined about his error in judgment, Bill was disgusted at the lack of efficiency, but Everett and Clay were as happy as poor kids peaking at a circus through holes in the tent.

After lunch I loaded the roundup pan, now filled with dirty dishes, stowed the leftovers, and lifted and locked in place the tailgate on the back of the chuckwagon that when let down formed a table on which to serve the food. We were, of course, taking the old road home.

Mike directed me onto the two-track lane, and as we zipped along, we both happily listened to the boy's excited visiting. It helped cover up an obvious fact: this road was so easily followed I didn't even need help finding the way home.

One fall when I was the cook at Babbitts, I was at Cedar Ranch, our summer camp, getting ready for the wagon to come back from the Catarac where Harvey's wife, Janet, cooked. When the chuckwagon arrived, I would load the groceries I had purchased into it and move it down to Spiderweb for a while and then move back up to Cedar Ranch and various other places on the COs for the fall works. I had to get us as a family packed up to stay away from home, get the groceries to feed a crew packed up ready to load, and be ready to put supper on the table shortly after we got down to Spiderweb. I was also hoping to leave my house clean enough to not be disgusting to come back to when I was tired but still in full-on cook mode. It was a busy morning.

I took the trash from the cookhouse out to the burn barrel which sat about four feet from the start of a huge wood pile that had logs cut by the Indian firewood-cutting-crew and an even bigger pile of old wooden posts that were one day going to be cut up for firewood, and a smaller pile of already split up cookstove wood. I started the trash burning and then studied the situation. I had no grate to put over the top, but the barrel wasn't very full, and the wind wasn't blowing hard, and I had so much I needed to do, and no one else ever worried about it. Why should I? I left to get other things done.

Not so much later, I went up the hill from our house to the backdoor of the cookhouse and saw smoke coming out from the middle of the unsplit log pile. I went closer and looked. It couldn't be—a tiny fire was burning right down at the bottom in the middle of the log pile. I couldn't have gotten those big logs to start if I had spent all day trying. I thought about throwing the logs that were on the outside of the pile away from the burning logs, but reconsidered because they looked too heavy and too many for that plan to work.

To make the perfect bad storm worse, that spring when Ed and I had moved to Cedar Ranch, and he turned on the water to our house that was always shut off and drained for the winter, the water had come into the line at barely more than a trickle, except in our bathtub. Even the faucet outside of house, not so far from the cookhouse, this year ran at just the smallest stream compared to all other years when it had gushed out stronger than town-pressure.

I got some buckets and strung the hose, which was connected to the outside faucet, uphill as close as I could stretch it toward the cookhouse and started it running in a bucket. I then put a bucket in the cookhouse kitchen sink and started it running. I then ran down the hill, through the house, and filled a bucket in the bathtub, and ran with it to the log pile. One bucket did an amazing amount of good. The smoke stopped for a second and then came back. I ran into the cookhouse. That bucket was only partly full, but I took it anyway and splashed it down into the logs. Again the fire slowed for a minute. Back into the cookhouse, start the water running into the bucket, run to the bucket under the outside faucet, run that small amount of water that had collected up to the fire, dump it on the fire which was winning the race and growing bigger.

The boys had tuned into the fact that their mother was stressed and doing weird things. I got another bucket for them, and told them to go to the horse trough, which was farther than all the rest of my water sources, and carry some water up to the fire. They were scared, so I went down with them and showed them how they could do it. I left them at that chore and ran into the house for another bucket from the bathtub and carried it to the fire. I was really losing. Everett and Clay showed up saying they could not do it, and they wanted only to cling to my legs and cry because they needed their dad to feel safe. I left them standing wherever they wanted to stand and continued my cycle of packing water from various faucets.

Sometime in the middle of all this chaos, the old, unhealthy waterman showed up, took one look at the situation, announced that he would go get the forest service at the edge of town 30 miles away, got out of his company pickup, got into his personal car, and drove away. The fact that he did not want to leave his personal car at the ranch by a burning woodpile, between three houses, a propane tank, and all the cowboys' personal pickups was not lost on me.

After a while I realized I was not going to win the battle so I just stood by the propane tank praying that it would not get so hot that it blew up. I told the boys to go stay down at the house, but they had to be with me, so I let them stand there, too. If that propane tank blew up, I wanted to blow up with it.

Bill Howell and his youngest son, Tom, were the first of the returning crew to show up. They were big eyed. Nothing needed to be said so I didn't say anything. Bill told Tom to go get the backhoe and push the wooden posts, which were now on fire, farther away from the propane tank. He walked up and touched the tank. I wondered if he knew anything about how hot a propane tank could be before blowing up. He asked me where the waterman was, and I told him in town getting the forest service. That wasn't lost on Bill either. He was also noticing that the fire had burned up high enough to burn the powerline in two that ran a well on top of the rim that supplied a great deal of the summer country with stock water.

The woodpile was all gone, the crisis was over. Ed showed up in the chuckwagon. I continued to stand there and rehearse and regret, so Bill told me in a few words to move on. Ed and I loaded the groceries I needed to feed the crew at Spiderweb, got our suitcases and the bedroll the boys shared, lifted Everett and Clay in, and headed down country to get supper on the table.

That night, laying on the floor in a hot house without a breath of a breeze, sharing a bedroll with Ed that he had been sleeping in while on the Catarac (not so clean), feeling the ache in my shoulders from trying to pack buckets of water at a run, wondering if I had done things differently if I could have put the fire out, thinking of the long day ahead of me cooking and caring for two busy boys—why couldn't I just let it go and fall to sleep?

The next morning when the alarm went off our routine pattern started clicking on schedule. Ed got up and headed to the bathroom to wash up, comb his hair, get dressed and step out of the door headed to the bunkhouse. As soon as he was on his way, I got up, made our bed, washed up, combed and braided my hair, got dressed, looked at the boys, committed them to Jesus' all powerful care, and stepped out into the darkness of a new day. It took less than a couple minutes to walk to the bunkhouse, but it was, as always, refreshing and peaceful. I didn't have to be alert for boogie men in the darkness because Ed had scared them away by going first.

As I stepped into the kitchen, Ed was just sitting down from building the fire in the cookstove and putting on two huge pots of coffee and a teapot of water. The woodstove top was getting warm, and I put the 16-inch skillet on it with a big glob of shortening to heat up, then got the steak out of the refrigerator and gathered the makings for the biscuits while keeping an eye on the oven thermometer mounted in the door of the wood cookstove, making sure it was rising and would be hot by the time I got the biscuits ready to put in the oven, while watching the skillet to be ready to put the steak in to fry when it got hot.

The coffee boiled, Ed got up and carried each pot to the sink and added a little cold water to settle the grounds and then made us each a cup with lots of canned (evaporated) milk and plenty of sugar. I never liked coffee except the coffee that Ed doctored up because he was generous with the cream and sugar. Fire in the woodpile yesterday or not, today was just another cooking day coming down.

After a few days at Spiderweb, we moved back to Cedar Ranch. The house I had worked so hard to leave clean had a dusting of ashes across the kitchen because somehow the door hadn't gotten shut correctly, and it had blown open and let the ash-filled breeze blow in. Oh well, there were no germs in ashes, and they would just have to lay where they were until I had the time to clean them up.

Everett and Clay took to trying to build an Indian fire in the ashes of the old woodpile. They had tried to build Indian fires for years, with my blessing, because I thought they had a zero percent chance of success. They got very dirty from crouching in the ashes, but they were right out the kitchen door of the cookhouse, which meant they were not stepping out in front of a truck or stepping on a rattlesnake in the tall grass. They stayed at it for days. We were all happy. End of story for me.

Everett clearly remembers that they did get a fire started by rubbing their charred sticks into dry grass that they had gathered. They had it going pretty good when he realized it was burning very close to the new little woodpile that had been brought in for me to use while cooking for the wagon's fall works. He roused Clay, and they got some buckets and hurried down to the water trough, filled their buckets, and doused the only Indian fire they ever got started without ever telling about their success until 40 years later.

Clay got all of that information mixed together in his little-boy memory and lived for 40 years thinking he and Everett had burned the woodpile down by starting an Indian fire. I don't know if being told the true story relieved or disappointed him.

Chapter Six

One day in 1986 the Babbitt crew was making a big drive in Mesa Butte Pasture, which is about 55 sections and has several canyons and mesas within its boundaries. We were gathering the whole thing with the center of the drive going up the canyon from Tommy Tank to Dust Bowl and on up the hill to the Tubs. Bill was leading the outside of the drive on the south side, and I was leading the drive on the north side, and there was eight or ten men in between us.

By the time the center of the drive got into the canyon halfway between the dirt tanks called Tommy Tank and Dust Bowl, a big herd of cows and calves had been accumulated; and the men in the center were having a tough time keeping cattle walking west toward Dust Bowl, a couple miles away, and then Tubs several miles farther.

I had been way out to the north as far as the reservation fence, which was three or four miles north of this main canyon the big herd was moving in. I had searched and found what few cattle were in that area. I had thrown those cattle south toward the center of the drive, handing them over to a cowboy I'll call Joe, who was flanking for me. He would then send them farther southwest to the next man, and everything was going fine.

Then when I reached a high spot not far from Dust Bowl I pulled up on a promontory where I could see a long way and realized that I was way out in the lead of the drags of the herd that was probably strung out for a mile and walking slowly along. I could see the drags about two miles away to the east, and I could tell that the several men there could use some help. There were more than two men between me and the drags, and I could see several of them. I made myself visible on my high spot in an attempt to send a signal that nobody needed to come and assist me because I was ahead and not having trouble. Good cowboys should have understood the mute sign language. I started back east in an attempt to push the men next to me toward the drags even though they were a long way off. Then, suddenly, Joe appeared out of a low spot that had kept him hidden and out of my sight for a spell. He loped up to me and started shooting the breeze.

Joe possessed the worst temper I had ever seen He would erupt into a screaming fit with very little provocation. He was very difficult to be around, and his temper made him a trouble causer; and, as far as I was concerned, he was a liability instead of an asset. But it was not my place to say. It was not me who hired or fired people.

I didn't want to talk to Joe, but, instead, I wanted to tell him to quit shootin' the breeze with me and go back to the east and help those other boys. I didn't need any help! But I thought if I said anything he was going to blow up and cuss me out, which wouldn't have bothered me any because I would have ignored him; but then, I knew from experience, he would be ugly to everyone, and the atmosphere would be dark the rest of the day, or perhaps several days. As he talked to me about nothing for 10 or 15 minutes, I was thinking to myself, *Why don't you just go back where you belong and make a hand!* I was in a real quandary about what to do.

Suddenly coming up out of a draw from the south came Bill Howell. He was riding a sorrel horse he called Bullet, and Bullet was moving fast, and he was lathered up. I knew from a long way off that Bill was mad. I knew why he was mad. When Joe saw Bill coming, he turned east and loped toward the drags. I stood on my high spot and waited for what was coming.

"What the hell are you doing? Ed, you know better than to sit here on your ass and watch those guys who are outnumbered while you are way ahead and not doing shit! What the hell is wrong with you? I thought you were a better hand than that," Bill said. He turned Bullet around and loped off whipping the horse down the hind leg.

Bill had never cussed me out like that. Actually, he had never chewed me out about much of anything. I didn't have it coming. I was sitting there because I was trying to preserve a little peace among Bill Howell's crew, because that hot-headed cowboy next

Chapter Six

to me would have went psycho mad if I had insinuated that he needed to go somewhere else. I was mad! Really mad! I didn't deserve a butt chewing.

I rode along and got madder with each step my horse took. I determined that I was going to have to quit. I didn't want to quit, but a man shouldn't have to take a cussing like that if he didn't deserve it. I got depressed. I was sick of moving my family all over from one job to another and never belonging anywhere. I had been moving my whole life. I was sick of being a *Grapes of Wrath* white-trash nobody. I was sick and tired of being sick and tired. I didn't want to have to quit, but I had to; a man shouldn't have to take a cussing he didn't deserve.

I rode along, and the drags finally reached Dust Bowl. We pushed them southwest toward the Tubs several miles away. The drive was pretty much thrown together now, and all the crew was riding and making a hand stringing the cows and calves out and keeping them walking. Bill was on the left point and way ahead and performing like the pro that he was. The cattle were moving better now, and things were going good. I was on the right point on the other side of the herd, and I was making a hand, but I was mad. Mad and depressed.

Suddenly out of nowhere a thought came to me, "What if you just rode up there and apologized to Bill? Told him you were in the wrong and would try to do better. Admit to him that he had a right to be mad."

I contemplated that for a moment then thought, *"Piss on him! I didn't do anything wrong!"*

And then a still, small voice said, "Go tell him he had a right to be mad, and you'll try to do better."

"Okay, I will, but then I'm going to quit. I've got to!"

I rode through the leaders of the herd that were now moving nicely. "Bill," I said, "I would like to talk to you for a minute."

"Okay," Bill said. His face was ashen. The color you get when your blood has been drained because of stress.

"You chewed my ass out back there, and I guess I had it comin'. That's fine. I want you to know that I can take an ass chewin' when I deserve it. I'm not mad at you and I'll try to do better. I did what I did because I was trying to preserve some peace in your crew. I'm not mad, but I do wonder what you would have done if you were in my place," I said. We were both sitting side by side, and our horses were not moving.

He looked at me for a long time and finally said, "I guess you were in a tough spot."

I looked back at him for several seconds and then I said, "Well, I'm not mad, Bill, and I will try to do better." I turned my bay horse around and went back to my place. The depression left me. I didn't feel like I needed to quit. From that day on until the day he died, Bill treated me like I was his best man. We never spoke another bad word between us.

Chapter Seven

When Clay was just three or four years old the cowboys starting noticing that he had a gift for recognizing horses. Every chance Clay and Everett got they would sit on the pole corral outside of our house at Cedar Ranch and watch the cowboys with the horses, especially roping horses in the evening. Each man would call out the name of the horse he wanted for the next day, and Bill Howell or Ed would catch and lead the horse out to him. A few of the horses were somewhat unique in their color, but most were regularly marked sorrels and bays; and then there were a few whose appearance was very similar to another horse.

It started with Clay asking Ed questions, "Dad, is that horse Blondie?" So the men started asking Clay questions. They started with the most easily recognized horses, and as he passed those tests, they moved on to the harder ones. He was really good.

As he got older and could get down and walk around in the corral, he was always being friendly with his horses, just spending time with them; and sometimes doing something someone else couldn't have gotten away with, but Clay could.

I wrote this story about Clay and horses years ago. One time when Valerie and Bill Owen were out at Spiderweb, I mentioned to her that I had written a story and sent it to the **Western Horseman** magazine for publication, and I had never heard back from them. She took a copy of it and said she had a contact and would see what she could

Chapter Seven

do. I told her not to pressure them to print it if they didn't think it was good enough.

Not too long passed before Valerie let me know that **Western Horseman** was going to print my story. I said, "You didn't pull strings, did you?" She assured me that she hadn't. I always have wondered if Valerie had her fingers crossed behind her back when she denied exerting any influence toward their positive decision.

"Mom, I want to send these pictures to the **Western Horseman**. They'll print them this time. They're good!" Clay's hopeful eyes looked at me then back to his most recent drawings. The pencil sketches were good for an eight-year-old with no noticeable talent. It was not a love of art that prompted Clay to draw, but a love for these, his only subjects, horses.

This evening as Dad and Clay's older brother, Everett, sat in the living room listening to Clay and Mom, a great idea occurred to Dad. "I've got it. I know how Clay can get his pictures in the magazine. He can ask for a pen pal. Dad was a genius. Visions of published pictures dominated all thought and caused a few negative facts never to cross our minds. Most obvious—Clay's hatred for writing made his penmanship unreadable, and more important—the Junior Artist and the pen pal sections were always separate in the magazine.

Happily Clay sat up to the table with school paper and pencil writing while Dad dictated the letter. It read, "I live on a ranch with 3000 cows and 100 head of horses. I would like a pen pal." No one noticed any missing information, and with one mind and heart, we turned our thoughts to the certain joyful conclusion.

A couple of months later one of the cowboys came home from town with the biggest news, "I see Clay's letter in this month's **Western Horseman** asking for a pen pal." Dad's idea had worked! Clay's pictures were finally published! A trip to town was hurriedly scheduled for the next day.

Before the school bell finished ringing, the doors burst open revealing two happy-faced boys. "Have you got the mail yet? Can we go right now?" Traffic across Flagstaff never seemed so slow or finding a parking place so difficult. But finally Dad was on his way into the ranch office to collect the mail, including the long-awaited magazine.

Dad's expression as he climbed in and passed the mail to Mom answered the whispered question, "Did it come? Did it come?" Two adults, two kids, winter coats and a month's worth of mail in a single cab pickup made it a struggle to open the magazine. Then, at last, there it was, the Junior Horseman section lay exposed for all to see.

Just a typewritten letter—no pictures. Stunned silence filled the vehicle. No attention was given to a letter with no pictures and no eight-year-old handwriting to tell the rest of the story until Mom glanced through the stack of mail. "Look at this. Three letters addressed to Clay."

Clay leaned forward out of his disappointed slouch and looked at his name on the envelopes. "Three letters. Read them to me, Mom."

All who wrote over the next few months assumed this person on the ranch was a man, not a boy; some assumed he was the owner of these mentioned horses and cattle. The letters came from all over the United States and even Australia. Mom read them all but gave up writing in return to try and avert hurt feelings at about the 20th and lost count around 60. Then the letter arrived from a young woman saying that the social prospects of where she was living were not working out. She could cook and she could clean. Could she come? No! Clay had all the cooking and cleaning help he needed in his mail-censoring mama.

Chapter Eight

In the spring, when Everett was eight years old, we were down in the Spiderweb roping arena having fun team roping steers: Bill, Vic, Karl Ely, Wes Keifer, Charles Kent, Everett, Clay and myself. Although we might enter a few other roping events, our main destination was always the Arizona Cowpunchers Reunion. We were all involved in that rodeo and looked forward to entering every year. Everett and Clay learned how to compete there, and it was a big part of their childhood.

Everett wanted to enter the steer riding at that rodeo, which was for boys 12 years old or younger. It was decided, by whom I don't remember, that Everett should get some practice, so Wes Keifer roped a big muley heifer by the neck, and Charles Kent caught it by one hind foot, and they stretched it out in the middle of the roping arena. While they held the 450 pound heifer down on her side, Everett took a long piggin' string and wrapped it around the heifer's chest like a bull rope. I helped him, and when we got all the slack pulled out of the string and his hand held snuggly in the makeshift handhold, we pulled the head rope off of her neck and let the heifer up, helping Everett stay secure until she was on her feet. We let Charles Kent's stiff nylon heel rope stay on her hind leg, no doubt thinking it would fall off within a stride or two. It did not.

The big red heifer took off jumping and bawling and fishtailing with Everett holding on to his makeshift bull rope. Then after making a few jumps, the heifer bucked Everett off, and he rolled down onto the ground and then up into a sitting position. The stiff heel rope, being still connected to a hind leg, was whipping and trailing behind. When the heifer had covered about 15 feet from Everett's landing spot, the rope whipped and curled around Everett's neck and face just as if a cowboy had purposely used Everett's head and neck like a saddle horn.

The heifer was gaining speed, and when the rope came tight it jerked Everett's head and body around spinning him like a top. It was ugly. The rope didn't just unravel itself; it sucked up tight and dug deep into Everett's hide. The tail of the rope ended up in Everett's mouth, and it hung and pulled and scraped. I thought his neck was surely broken.

When the rope finally came off, Everett had a third-degree burn all around his throat and leading up into his mouth. The burn was deep and instantly turned as red as ketchup. He was just at the age that his baby teeth were going to fall out, and four of them, on top in front, were now gone. One of them dangled out of his mouth by a piece of skin. His mouth was bleeding.

When all of the stress and turmoil and commotion died down, we decided that Everett wasn't going to die. The teeth were about to come out anyway, and the third-degree burn around his neck would heal. Probably. We went to the saddle house and unsaddled our horses and went to the house.

Several days later Everett was at school playing with his friends out on the playground during school recess. A young woman walked up and started watching Everett. His neck looked worse than most men who have miraculously lived through a mob lynching. His cheek was swollen and bruised, and his gums were bloody and swollen. She stared, thinking to herself.

She asked Everett to come with her, and she led him over to another woman who was also a school official but of higher rank. "So what happened to your neck? How did you get those bruises and burns? What happened to your mouth?" the woman asked him with a very official and concerned tone in her voice.

"I got bucked off of a heifer," Everett announced proudly.

"You got bucked off of a heifer? What do you mean?" The woman's eyes grew wide.

"I was getting practiced up for the steer riding at the cowboy reunion," Everett said, glad that he could be counted among the top hands who had suffered for their craft.

"You were riding a cow? Were you by yourself?"

"Oh, no, my dad was there, and Charles and Wes and Karl."

"So did your father make you get on that cow?" the woman asked.

"No, I wanted to get on the heifer. I wanted to practice," Everett answered.

"So you weren't told to get on that animal?" she asked.

Everett realized the women were motivated by some kind of an agenda. He smelled a fish even though he was only eight years old. "Nobody made me get on the heifer. I wanted to get on and practice. Nobody made me."

The two women looked at him with cocked eyebrows and stern looks on their faces. Finally they released him, and he walked away with them following after him and wondering what a strange thing it must be to live on a ranch out in the middle of nowhere. You might just as well live on Mars.

When I was a kid, or even a young man, cowboy boots with pointed toes were very much in style. My first pair of made-to-order boots were Paul Bonds with pointed toes, but, alas, times and styles change even if parents refuse to go along with the progress.

By the time Everett and Clay grew big enough to care about such things, George Strait, Jake Barnes and Clay O'Brien Cooper had reinvented the wheel and changed the world forever. George Strait's hat always had a flat, or near flat, brim with sides that turned up just enough to distinguish himself from an Elko County buckaroo but was definitely shapeless compared to a dedicated Texas puncher with the traditional taco-crease. Jake and Clay wore Justin Roper boots with wide, round toes, and heels lower than Keds tennis shoes. George Strait followed suit with his boots because his biggest dream was to be a well-known team roper. So to my kids, especially Everett, high-heeled cowboy boots with pointed toes and cowboy hats with some old-time shape were definitely out, at least he hoped so.

There was a western wear store in Flagstaff known as Gene's Shoe Hospital. One day I was in there looking around through their hundreds of available styles of cowboy boots, and I came upon some kid's cowboy boots that were actually very well made. They had genuine leather soles, pegged shanks with steel in them, and underslung cowboy heels. They also had pointed toes complete with a beautifully stitched toe bug, just like real boots are supposed to have. And they were cheap, I mean they were actually cheap enough that a cowboy making $700 a month could afford them. I couldn't believe it! I bought two pair, one for each of the boys, and was sure my boys would be loving their dad for the rest of their lives for being so good to them.

I got home and had the boys try them on. I was jealous and wished the store had some like them in my size. I would have bought three pairs! The boots fit perfect. I was so happy and proud of myself for fixing my kids up with such quality stuff. I don't know if Clay cared, but Everett immediately reacted negatively. I asked him what was wrong.

"They've got pointed toes." Everett said.

"So What?"

"I don't like pointed toes."

"You don't like pointed toes?" I asked. "What the heck are you talking about? Look at the shanks on those boots." I pointed to the bottom of the soles, "Look at those pegs hammered in there! Those are real cowboy boots!"

"I'd rather have Justin Ropers," Everett said. Clay gazed out the window and watched a cottontail rabbit scurry by.

"Everett!" I said. "These are real cowboy boots; I mean, I wish they could have had some that fit me. I would have bought them."

The boots remained a point of contention until they wore out. I could never understand it. There were other things I didn't understand.

Both my kids had to grow up using worn-out nylon ropes. I've seen lots of team ropers throw down a new rope that had been used to catch maybe three or four steers, but because they had missed a steer or two they blamed their rope and would then throw it in the discard pile, that probably had several dozen more just like it, and then they grab a brand new rope and back in the box and rope another steer. That type of behavior never happened around our place because we couldn't afford a new nylon rope every time we missed a steer. When we were through with a rope, it was ruined. Completely worn out!

Craig Hamilton came to Spiderweb and put on a team roping clinic, and Everett happened to have a new rope, but it had a pretty bad backswing in it. Team ropers will know what I'm talking about. Craig took Everett aside and showed him how to turn his loop backwards one turn, and then when it is fed some more rope into the loop as you swing it, the backswing comes out of it. Again team ropers will understand. Well, all old ropers have tried this trick from time to time with varying degrees of success, but nobody puts a backward spin on their loop every time. Doing that will do some weird things to a nylon rope.

Everett's rope worked better after administering this trick, but instead of doing it a time or two and then quitting the process, he continued doing it. Pretty soon his new rope had some heavy-duty black marks on it from dallying on some big steers while having the saddle horn wrapped with inner tube. After multiple steers being caught and the backward spin being administered every time, his rope was not only black, it also possessed some deadly looking kinks in between his hands. *My word*, I thought to myself, *my kid's going to get one of those weird coils sucked into a dally along with his arm, and his arm will get cut completely off.* Most team ropers cut thumbs off, but the Ashurst family is going to start cutting larger more important appendages off. Nothing moderate about us.

"Everett," I said, "you need to quit putting that backward twist in your loop. You are going to get hurt."

"Craig told me to do it," he said defiantly. "And it's helping my roping."

Well, Craig had competed at the National Finals Rodeo, and I had never done anything but catch a few wild cows. I didn't know anything, especially compared to Craig Hamilton. He even wore Justin Roper boots and a straw hat.

Because of some heavy-duty divine intervention, we managed to not sever any body parts.

About this time we bought Everett a new Resistol black hat. I think it had a little shape in it when it was new. He administrated some steam to it and pretty soon the brim was flat. It was even too flat for George Strait. We, Everett and I, went around and around about that hat. I couldn't stand it. I couldn't stand my kid looking like an Amish steel wheeler so I confiscated the hat and put some shape to it. He was mad.

Soon afterward we were loading up into our vehicle and going somewhere. We were all dressed up. When we got home, Everett got out of the vehicle first and he held the car door open for Jean Ann. He had on his new black hat that I had just put some shape into the brim. Jean Ann was in a playful mood and stuck out her hand and said, "Jean Ann Ashurst." Everett took her hand and said, "Shell, Taco Shell."

Chapter Nine

Harvey Howell is Jean Ann's third brother, being ten years younger than Bill and seven years older than me. Harvey worked for the Babbitt Ranch for 30 or more years. When I first went to work for Babbitts, Harvey was still commonly known as Boog, a nickname he had acquired when he was a boy. I had met Harvey several times before I went to work for the outfit and had only known him as Boog.

Harvey was my friend. When I first went to work for the outfit, he held the position known as the jigger boss, which meant he was second in command, or what some old-times referred to as the pissin' post. You could have the glory of giving some orders, but, when push came to shove, you really didn't have any authority. If things went well, it was because of the wagon boss' expertise, but if they went bad, it was because of the jigger boss' incompetence; or at least that is how some people viewed the position. Harvey filled the position well, and in 1974 he and his brother Bill, who really ran the outfit, got along real well. Harvey and I hit it off right from the start. He loved to rope, and he was an outstanding roper. If I had a dollar for every cow I roped with Harvey I would be a multi-millionaire.

When I went to work for the outfit in May of 1974, Harvey and Janet had only been married a few months. About a year later on March 11, 1975, Harvey came home from the Flagstaff hospital and met me at the front door of the Spiderweb bunkhouse and announced, "Well, Ed, I had a son! Todd Neil. Yep, that's what we're going to call him, Todd Neil Howell." He was excited! It was amusing how he related to me that he had just had a son, never mentioning Janet, as if she had not participated in any way.

Todd was a pistol right from the get-go. He was always moving forward—full blast. When Todd was about four, Harvey and his family moved to Redlands, which is a camp on Babbitt's ranch 80 or so miles west of Spiderweb but still part of the same outfit. He was always doing something with great enthusiasm.

There was a perfectly good outhouse on the backside of the corrals at the Redlands. A good usable outhouse is a fine thing to possess on a big cow outfit for obvious reasons. One day some years later, Jean Ann and I drove over to visit Harvey and Janet at Redlands. We were living at Spiderweb at the time. In the middle of the day, for some reason, I ventured out to the corrals and heard a considerable amount of banging and pounding taking place somewhere off in the distance. I went searching for the source of the noise, and I found Todd with a large hammer and crowbar, and he was dismantling the outhouse. He had enlisted Everett, who was about five years old and two and a half years younger, in his demolishing project and they both had found quite a few tools at the shop to assist them in the work. "What are you doing, Todd?" I demanded.

"Dad wants this outhouse tore down, so we're tearing it down." Todd answered.

Harvey did not want the outhouse tore down, but it was too late. It was demolished.

In the summer of 1981, we went to the Cowpunchers Reunion, which was held that year for the last time at John Avery's Kowboy Kountry Klub on the north side of Flagstaff. The reunion and rodeo was small in those days and for the most part attended by bonafide ranch folks. If you won first in an event, your winnings would have barely paid your expenses for the weekend, especially if you stayed in a motel in town; but most folks just camped out among the pine trees, sleeping in teepee tents or gooseneck stock trailers that had been swept out, which removed most of the fresh horse manure. Jean Ann and I lived north of Flagstaff 35 miles at Cedar Ranch, and we just drove back and forth.

Todd was six years old that summer, and he and Cole Gould, Charlie's boy, struck up a friendship and played all weekend with Everett, Clay, and some other kids in tow; but since Todd and Cole were older they were the big bosses of the gang of kids. They ran and played among the pine trees and corrals and acted like wild Indians, which is what little boys that age are supposed to do.

Finally, after performing every other heroic act they could think up, which usually resulted in the suffering of some younger kid like Everett or Clay, Todd and Cole decided that one of them should arrest and tie up the other one to a pine tree for some heinous deed such as horse theft. Since Todd was bigger than everyone else, including Cole, he became the one doing the tying, and Cole became the one being tied. It had all been fun, especially for the two of them, until this new development. Todd overpowered Cole by sheer size and weight, and with an old nylon rope, he lashed Cole to a pine tree while Everett looked on with great trepidation.

Cole got mad, whereas Todd wasn't mad, he was just big; big and having fun. Cole's face turned red and he began swelling up with air, obviously preparing for a delivery of threats and predictions of retribution as Todd kept tying nylon rope. By instinct Everett knew that Cole was about to exhale with an onslaught of the severest profanity ever to come forth out of a child. "You! You! You! You old poop eye!" Cole finally screamed.

Well that was about as unchristian a statement as Todd had ever heard. He could tell that Cole was truly mad, so he untied him; and within a minute or two, Todd's exhibition of bullying had been forgiven and the two of them became friends again and went to chasing the younger boys through the pine trees.

In the fall of 1982, all of the Babbitt Ranch calves got sick with some kind of respiratory ailment. Most years there wouldn't be one sick calf on the whole ranch, but, for some reason, an epidemic of biblical proportions went through the calves after they were weaned. The calves were all turned out in several pastures that were 70 or 80 square miles in size and when the sickness hit it went through the cattle like wild fire. We were totally unprepared for it. In those days real good medicine wasn't available, at least as good as today. At first we didn't even have any medicine because most years nothing ever got sick. Later when we got some medicine we were all doing lots of extra riding, trying to keep up with doctoring sick cattle, and it was already a very busy time of year trying to finish the fall roundup.

I had broke a sorrel horse to ride that was called Swampy. Harvey had ended up with Swampy, and he had given him to Todd who was seven years old. Harvey would take Todd with the crew working cattle, including many extra hours covering huge expanses of country looking for and doctoring sick calves. Many days Todd would ride from sunup to sundown on old

Swampy, and at that age he would do his share of roping the sick cattle. On several occasions I went home at night and told Jean Ann, "That darn Harvey is going to kill that kid. He's taking him all day long and working the daylights out of him! He needs to give the kid a rest." But Harvey kept taking Todd, who never weakened or quit grinning from ear to ear. He just kept whipping old Swampy down the hind leg and swinging his rope. He was one tough kid.

In the spring of 1983 we were branding calves on the W Triangle Ranch, which is Babbitts' outfit on the west side of Highway 64 which goes north from Williams to the Grand Canyon. Todd was helping and was seven years old, and he was riding Swampy. One day we made a big drive into a dirt tank called Red Dyke and branded calves there, probably about 200 head. When we were through and had turned the cattle loose we proceeded to load our horses in several bobtailed trucks we had there. Todd got up inside one of the trucks ahead of his horse. When Swampy jumped up and into the truck, Todd failed to step aside, and Swampy jumped on Todd and broke his leg really bad.

We got Todd out from under Swampy, got him loaded into a pickup truck, and started driving toward the Redlands where Harvey and Janet lived. There was also a bunkhouse there, and the branding crew was staying in the bunkhouse. I was driving the pickup, and Harvey was riding shotgun holding Todd who was trying to act brave even though he was in a lot of pain. Finally as the pain got unbearable, Todd decided he would go ahead and die. He looked at me and said, "Ed, you can have my saddle," and, after pausing for a moment, he looked at Harvey and said, "Dad, you can have my bed."

"Ok, son," Harvey said softly, although there was a slight grin on his face. "Thank you."

Todd was quite a roper when he was a kid. When he got to high school he added bull dogging to his repertoire. There was a kid who lived in Flagstaff named Cody Hart who was Todd's age and a real good kid. He was going to the high school rodeos at the same time as Todd, and he had a bulldogging horse.

One day we had a big practice session at the roping arena at Spiderweb. Harvey and his family came over from Redlands, and Cody Hart came out from town and brought his bulldogging horse. Everett and Clay were mounted on their rope horses, and we had a big time practicing all of the timed events. But neither Cody nor

Todd had ever had much experience bulldogging, and they were riding past the steers without getting down. Time after time they failed to take the plunge and get down but, instead, would gallop on by.

I had done a little bulldogging 20 years earlier but hadn't got down on a steer in about that long. Finally I told Cody to let me on his dogging horse so I could show them how to do it, and I did just that. There was nothing fancy about it, but I did make a successful run including having my feet land out in front of me enabling me to twist the steer down. But to no avail. The boys still wouldn't get down.

Harvey got mad and told Todd to, "Get on that horse! And you are going to get down!" Harvey got Todd hazed down the pipe fence on the backside of the arena and whipped the horse Todd was on into a dead run. "Now, get down! Lean over and rest your arm on that pipe and get down on it!" Harvey yelled as he was whipping Todd's horse. I think he was whipping Todd also. Todd leaned over off of the right side of his horse and laid his arm over the second pipe from the top rail on a pipe fence that had upright pipe posts every 15 feet or so. These posts and the pipe rails were all welded together. So at a high rate of speed Todd hooked his arm across the top of the second rail and slid along it for a ways with everything going fine for a second or two. Then Todd's arm and shoulder came to an upright post that was cemented into the ground and welded to the horizontal rail. Todd came off—really fast.

Cody Hart watched that and realized that Harvey was staring at him, and so Cody asked if someone would load another steer into the chute, and right then he started getting off on the steers' backs. He didn't want anything to do with Harvey's training technique using the pipe fence.

Chapter Ten

Bill Howell started putting together a herd of his own cows which he ran on several different pastures he leased around Northern Arizona. He continued to run the Babbitt Ranch while he did this, so his own cattle business was a sideline, or sort of a sideline, perhaps it was moonlighting, or an extracurricular activity. He had country leased from a ranch north of Williams and bordering the Grand Canyon National Park to another piece of country near Winslow and east of Flagstaff 50 miles. Bill's good friend Doy Reidhead named Bill's cattle venture Badly Scattered Cattle Company. From the first time Doy said those words, which were meant to be a joke, Bill's cow-doings were referred to as Badly Scattered.

When Everett got to be 9 or 10 years old, Bill started taking him on his extracurricular trips to work Badly Scattered cattle. It wasn't long before Clay was going also. The two of them, Everett and Clay, had been helping me and the rest of the Babbitt crew with cow work for several years and could be a lot of help at times. They certainly could get behind a bunch of cows and trail them for miles or help pen them in a corral or work a gate if someone was trying to do some sorting. They had saddles, ropes, and cowboy hats and were wanting to grow up and become cowboys. Going with Uncle Bill and helping him do something with a bunch of Badly Scattered cows was a lot better than watching cartoons or riding a skateboard.

It would take several hours of driving to go from Spiderweb to one of Bill's Winslow leases, and most of the time Bill would drive along in silence, lost in his own thoughts. Everett was always nervous and sat straight up and would be paying attention, whereas Clay, who was seldom nervous about anything, would fall asleep and be relaxed the whole time. Sometimes Bill would talk, and because of old age he would get to mumbling and talking softly to the point nothing he said made any sense, but Everett would make sure he nodded and said yeah or uh huh occasionally so Bill wouldn't think he wasn't paying attention.

On one trip to Winslow, Bill and the two boys got a late start so when they passed the old Twin Arrows Trading Post, 20 miles east of Flagstaff, Bill stopped at the joint to buy the three of them lunch. This was a big deal because Bill was tight with his money and didn't very often spend money on frivolities such as food.

The old Twin Arrows truck stop was a sure enough greasy spoon outfit and very seldom got any business. It had an old lunch counter with round stools covered with red naugahyde. There were no tables, just the lunch counter, and so the three of them—Bill, Everett and Clay—sat on a round stool at the counter and were waited on by a fat cook wearing a greasy tee shirt and a hat that was even greasier. He had tattoos and several days' worth of sweat all over. Bill ordered a cheeseburger, so Clay ordered a cheeseburger, but Everett looked at the menu and decided on a hot dog. Bill gave him a funny look but said nothing. The greasy cook put a couple hamburger patties on a big griddle, which was just several feet on the other side of the counter from where they all were seated. Then he got in a freezer and took out a frozen weiner and dropped it into a pan of cold water that he put on top of a propane burner turned down low.

The cowboys sat as the greasy cook watched the hamburgers cook with his back turned away from them. There were flies everywhere. The cook started swatting flies with the spatula that he was using to turn the hamburger patties with. When he was successful he would scrape the spatula against the griddle, thus depositing the squashed fly onto the greasy griddle top. The two Ashurst boys watched as Bill sat grinning at them. When the hamburgers were cooked the greasy cook put the meat between two pieces of bread and sat them in front of Bill and Clay. Then he rescued the half frozen weiner out of the tepid water and put it in a hot dog bun and sat it in front of Everett. The weiner was cold, but Everett ate it anyway. Everett knew that food cost money, Bill's money; and he was acutely aware of the fact that he had made the decision to be different when he could have ordered and received a hot, greasy burger fried with fly guts. He knew that Uncle Bill didn't approve of complainers.

Chapter Ten

Occasionally, when driving along, the overly conservative Bill would offer the boys a piece of chewing gum. He would take a stick of Wrigley's spearmint gum and divide it into two pieces: one for Everett and one for himself. If Clay was with them, he would split it three ways, a third of a stick for each cowboy.

One fall when it was time for Bill to ship Badly Scattered Cattle Company calves he took several of the Babbitt cowboy crew with him to help him ship cattle off of the Red Gap Ranch near Winslow. I was part of the crew that day, as was Everett who was ten at the time, and Clay who was nine years old. We had been working cattle at Babbitts every day for several months with little or no time off, so we were all well worn. Bill and his brother Harvey were sort of partners on the cows. Although none of us really knew what their partnership agreements were, we were aware that their relationship was somewhat strained. But it was none of our business, and Bill made sure it stayed that way.

The corrals at the Red Gap Ranch were very poorly made and even more poorly designed and Badly Scattered Cattle Company was only leasing the outfit so very little money was spent on improvements. There was an old wooden gate down on one end of the set of corrals that was in need of repairs. Bill had brought along one new 2X10 board to repair the gate. Late in the day he told Harvey, "Take these boys and go down there and fix that gate. I've got a board in my pickup to fix it with."

The boys meant Everett and Clay, and so the three of them took off walking toward the gate that was sagging and dragging the ground to the point it had to be packed by a man when opened or shut. The top board was broken in two, making the gate less than four-foot tall. Harvey, who was always chipper and happy-go-lucky acting, retrieved the new board and packed it to the sickly gate; and he talked to Everett and Clay as he walked along.

"Well, I wonder how we are going to fix this gate," he said looking at his two nephews as he bounced along. He instructed them to hold one end of the board up against the old gate as he held the other, and he took an old rusty nail and used it to mark the spot where the board needed to be cut. "Well, boys, I wonder what we are going to use to cut this board?"

The three of them walked over to Bill's pickup and could not find a saw, but they did find a large digging bar with a pointed end, such as you would use to dig a post hole in rocky ground. They also found a shaping hammer like you would use to shape a horseshoe on an anvil, and they found a coffee can of bent, rusty nails. They walked back to the sickly gate and the new piece of 2X10 lumber. "You boys sit on that end of the board," Harvey said as they laid the board flat on the ground and then Everett and Clay sat on one end while Harvey started plummeting the beautiful piece of lumber with the end of the digging bar, smashing into the board with the dull bar at the very spot where he had made the mark with the rusty nail. When the new board had been partially bludgeoned in two, Harvey looked at his companions and said, "The Anasazi always get their man!" He then crushed the board in two, and with the boys help they nailed it in place at the top of the gate.

One time the boys were at Red Gap with all the Howell crew: Bill, Harvey, Vic, Todd, and Dallas. The boys were about nine and ten years old at the time. They had gathered a pasture and penned the cows in a barbed-wire waterlot that was pretty dilapidated. Off in one corner there was a smaller corral that was also greatly diminished in its ability to hold cattle securely because of its old age and lack of maintenance. The entire setup was very questionable as far as holding a herd of cattle. Then to make matters worse, some of the cows were crossbred cows with a little brahma blood that Bill had bought from his friend Doy Reidhead. Bill didn't like brahma-cross cows, but they had been brought cheap, which made them acceptable. Doy's cattle tended to be a little high headed and had been well introduced to loco weed, so they were skittish if not downright crazy.

The cattle were successfully penned in the big wire water lot, and then successfully put in the smaller corral in a corner of the bigger corral. Everett, Clay, and Dallas had been told to go off to the side and set up the branding outfit which consisted of a big pipe with a hole in the side and a burner connected to a hose that was subsequently screwed into a five-gallon propane tank. The three smaller boys had been told to get all of this stuff ready but had not been given permission to start the fire for fear they might burn the ranch down. Actually, there was nothing present to burn except red sand and caliche, but they weren't supposed to burn it up. Bill had ordered Cousin Todd to get off his horse and open and shut an old wooden gate that needed to be picked up and carried if someone wanted to move it. Bill was going to cut some cows out the gate that Todd was trying to manhandle. While all of this was going on, Harvey and Vic had stopped close to the first gate that the cattle had passed through, and they had their tally books out and were arguing about a discrepancy in the cow count. Beings Harvey and Vic both had a small share in the cowherd, whereas Bill had the larger share, a discrepancy in the cow count could be a sticky subject. Anyone who ever witnessed family partnerships in the cow business will understand.

The three small boys got the branding outfit set up, minus the flame; and then Clay and Dallas began playing in the sand with their hands while their horses were tied to the fence nearby. Everett, who was always standing at attention, was sitting on his horse nervously awaiting an order to do something, and Harvey and Vic were setting on their horses arguing. Bill was riding a big bay horse he called Jeb and he was cutting cows out the gate that Todd was working afoot. Jeb had a habit of making Bill upset, and the more Bill got upset the redder and hotter the flames became that were coming out of Jeb's hind end. Bill, who was a master at working cattle, was trying to be calm; but Jeb was making him mad, and he began jerking and jobbing, if you know what I mean. The Doy Reidhead cows were starting to look at the low-hanging top wires on the corral fence and were dreaming of jumping out. Bill would cut a cow out of the gate and then jerk on Jeb who would back up and smash twelve-year-old Todd against the fence and thus pin him and make it impossible for him to shut the gate, which would turn unwanted cattle out of the gate, which would make Bill madder; and he would jerk on Jeb and back him up and smash Todd again. Todd's face was shoved into close proximity of Jeb's posterior many times.

And then it happened. An old black gentle cow spied the gate that Harvey and Vic should have shut, but they failed to do so because they were making sure no one was going to be cheated out of a cow. The cow started walking toward the open gate. Harvey and Vic were too busy to notice, Everett noticed but it was really Harvey or Vic's job to stop the cow, so he was frozen in indecision and nervously did nothing. Clay and Dallas visited and played in the sand. Vic and Harvey argued instead of paying attention, and Cousin Todd had his head shoved up Jeb's hind end; but Bill, who never missed anything, saw the wreck that was about to commence: An old gentle Angus cow was about to walk out of an open gate.

Bill spurred Jeb who immediately expulsed a flame akin to the fire coming out of an F-16 tail and stampeded toward the gate but made sure Jeb's feet threw sand on Clay and Dallas as he flew past them screaming at them, "I hate visiting! I hate visiting! Quit your damn visiting! Visiting and smokin' cigarettes! I hate visiting!" He flew to the open gate and made a hand keeping the old Angus cow from getting away. Everett was too scared and nervous to do anything but was not guilty of that wicked visiting like Clay and Dallas. Harvey and Vic kept waving their tally books and were oblivious of Bill's meltdown, and when Bill got the cow run off that Vic and Harvey should have turned he went back to cutting cows out of the gate and smashing Todd against the fence with Jeb's fire-bearing posterior. Clay and Dallas went back to visiting and playing in the sand.

One time when Bill and Everett were driving home and were about 20 miles east of Flagstaff, Everett had to pee so bad he thought he was going to lose control of his bladder, but he said nothing. Abruptly Bill asked him, "What's the matter with you, kid? You gotta' pee?"

"Yeah," Everett said.

Bill immediately turned his pickup off of the east-bound lane of I-40 doing about 70 miles an hour. Everett thought they might wreck and the stress of the landing almost made him pee his pants. "Nobody should ever have to go through that much pain!" Bill declared as he skidded the pickup to a stop, allowing the 10-year-old to jump out and pee. He could be gruff and then at other times he would be gentle.

One spring when Clay was eight he got to go with the crew and camp at Redlands, which was southwest of the Grand Canyon Village 35 miles. There were 10 or 12 mature men including myself, and we busied ourselves branding calves every day for about eight days. We penned a big bunch of cows and calves near Redlands and branded the calves. During the process, two small calves got out of the branding corral but stayed close by. They were not really stirred up. Clay was riding Waco who was slow and dumb, but Clay had managed to rope several slow steers on the horse, and he was sure he was a top cowboy even though he was eight years old. Clay decided it would be best if he pursued the calves and roped them, and he immediately ran one of them through a fence, which created an even bigger problem. With that done he took off after the other, having a lot more open ground to work with on the second one. He ran him several hundred yards, and the calf crashed through a fence and ran off, creating more problems.

This type of behavior was definitely frowned upon by management, as well as the cowboys. You weren't supposed to be running calves off and through fences. You were supposed to catch them smoothly without creating a mess. You weren't supposed to act like a gunsel. Instead of screaming and hollering and cussing Clay out, Bill rode up to him, and in a conversational tone of voice he said, "You know, sometimes a fella might do things differently, like you could have used all the room you had to turn that calf away from the fence rather than run him straight toward it. You know that might have worked. It would have given you a lot more room to work with. You might think about that next time." That's all that was said.

Chapter Eleven

At Cedar Ranch there was an old shop building on the hill above the house. It was the first building that you drove by when you drove into the ranch and when you passed it you went downhill toward the barn, horse corral and the several other dwellings that were there.

The old shop building was similar to a lot of other old shop buildings on many ranches scattered over the West. Back in its dark corners and unswept benches and cabinets were a multitude of old pipe fittings, bolts, rolls of bailing wire and things like oil filters, some of them used and some not, plus various archaic tools including several old shovels with broken handles.

Everett and Clay were very young. We didn't have a yard fence, and one day they wandered up to the old shop building and went to exploring around through the grease and dust and found a box of peculiar-looking things that had been forgotten and thrown into a corner where cobwebs had grown over and helped to hide the contents of the bullet-looking steel things. They filled their pockets with the little metal cylinders and were heading down the hill not knowing what they might do with their new find. A maintenance man happened to drive up as the two little boys were leaving the building. "What are you boys doing?" the man asked.

"Oh, we're just playing," one of the boys replied.

"What have you got stuffed into your pockets?"

"We don't know what these things are." Everett took one out and showed it to the man.

"Let me see." The man's eyes got real big. "Okay, now take those things out of your pockets and hand them to me. Real careful." They were blasting caps for dynamite; any one of which had the power to blow a kid's leg off. And they had handfuls of them in their pockets.

Years later the boys watched the movie *The Man from Snowy River*, and like every other male healthy enough to breath, the boys were fascinated with the cracking of bullwhips while chasing herds of wild horses through the wilderness. In a ranch shop building at Spiderweb, the boys got to looking around and found some neat-looking plastic cord that was rolled up and thrown into a corner. They cut a couple five-foot lengths of the orange-colored cord and tied both pieces of it to some long sticks and made themselves a couple bullwhips just like the man from Snowy River. Man this was neat, they thought, because when they cracked their whips, sparks started flying. They were instant bullwhip artists.

The orange cord was primer cord used to ignite dynamite. It is a plastic hollow tube, similar to a flexible straw, and filled with explosive; each four inches of which would have the equivalent explosive power as a number 8 blasting cap. No wonder their whips made sparks when they cracked them.

Chapter Twelve

Everett and Clay rode a school bus from Spiderweb to school in Flagstaff for eight years. The bus would run from the Weitzel grade school and Coconino high school, both of which were located in East Flagstaff, all the way out to the town of Gray Mountain on the Navajo reservation, which is a total of 45 miles. Spiderweb on the Babbitt Ranch is seven miles south of Gray Mountain. This bus route traversed the summit at 7250 feet in elevation, about 10 miles north of Flagstaff. After going over the summit and heading north, the highway dropped a couple thousand feet in elevation before reaching Spiderweb and a couple hundred feet lower at Gray Mountain.

When Everett was five and Clay four we had a slight accident near Rawlins, Wyoming when our Ford pickup spun out of control on the freeway during a blizzard. This little mishap was inconsequential as far as damages go but established a fear of icy roads in Everett's brain.

The Gray Mountain bus would inevitably have to drive on icy highways and streets during the winter, and for several years these icy roads caused a great deal of anguish for Everett, and perhaps Clay also; but he was easier going and didn't say as much about the ice as Everett did.

When Everett was in the fourth grade a winter storm moved in and began dropping big wet snowflakes, and Everett sat in his fourth grade classroom watching the snow fall and becoming more apprehensive as the hours of the day clicked by. By the time it was bus-loading time, the roads were packed with wet snow, and it was white and icy looking everywhere.

Everett and Clay sat up in the front of the bus right next to the bus driver and were as rigid as icicles when the bus pulled away from the elementary school. They stopped at the junior high and loaded up some more kids without mishap and then went on to the last stop, which was Coconino high school. As the bus pulled away from Coconino high someone pulled out in front of the bus causing the driver to quickly apply the brakes, which in turn put the bus into a slight skid. Everett was hair-triggered after living in dread all day long watching the snow fall as he looked out of the school window, and as the bus slid to a stop on the ice, he got up and told the bus driver, "Open the door. I am not riding this bus home and neither is my brother."

The bus driver's eyes got big as he looked at this 10-year-old kid with wild eyes and a determined look on his face. Something not too far short of an international incident began to unfold. You can't just let kids off the bus anywhere. What was he supposed to do? Pretty soon a school counselor and the principle were involved. Everett held his ground; He and his little brother were not staying on that bus!

Bill Howell's stepdaughter, Helena, was also riding the bus. She was several years older than Everett and considerably more calm. She told the authorities that if they would allow her to use a telephone she would call her mother, who was working in town and not too far away. Her mother could drive all of them home as soon as she got off of work. And so all of that was arranged, and Linda Howell drove Everett, Clay and Helena home, and so they all arrived safely, although a little later than usual.

The next morning Jean Ann and I wondered if the trauma felt by Everett because of the icy roads and sliding school bus was lessened after a night's sleep or were he and Clay stressed out and worried about having to keep getting on the bus. After talking to Everett it was apparent that he was more stressed than we had realized and riding the bus on icy roads terrified him even though he had been keeping it to himself.

I wondered what to do, and the thought came to me to read him a scripture out of the Bible. I chose Luke chapter 12 verses 6 and 7, "Are not five sparrows sold for two cents? Yet not one of them is forgotten before God. Indeed, the very hairs of your head are all numbered. Do not fear, you are more valuable than many sparrows."

These words seemed to give Everett peace, and later he said that he didn't worry about the bus and icy roads ever again. But there was more trouble ahead, and Everett played the leader and protector over Clay, and they stuck together.

A year later the kids loaded up on the bus at the grade school and then proceeded on to Coconino high school to pick up the high school kids. It was a cold winter day, and the roads were very slick and icy. After leaving Coconino high school, the bus took off on icy streets headed toward Santa Fe Avenue. Several intersections past Coconino high, as the bus was going through an intersection, a big dually truck came from a side street and couldn't get stopped at the intersection; and with all wheels locked up, the big dually truck came skidding into the intersection and hit the bus on a front fender with its front bumper and knocked the bus sideways, and it skidded across the ice at a 40 degree angle.

The bus driver at the time was the best driver that ever drove while the boys rode the bus. She was a skinny little old lady with gray hair, a two-pack a day cigarette habit, and weighed about 95 pounds, but she was a good hand. The collision was pretty brutal due to the weight of the heavy dually truck that crashed into the bus' front end. It separated the lady bus driver from her seat even though she had her seat belt on, she being so skinny there was plenty of slack between the seat belt and the chair she was supposed to be strapped into. She was knocked upside down with her head coming to rest on the floorboard and her body being twisted, but somehow she held a foot on the brake pedal. The impact broke her wrist, but even then she kept her little hands gripped to the steering wheel while her head rested on the floor. Everett and Clay were sitting on the front row, and Everett went to her as she dangled in a very contorted position, "What do you want me to do?" Everett asked her.

"Put the transmission in park," she replied as she dangled upside down, and so he did.

Off to the side of the road and directly in the middle of the bus' new trajectory was a ditch made by some new construction, and if the bus had slid another three feet it would have certainly turned over on its side and slid down the embankment.

Everett unbuckled the little bus driver's seat belt, and she finished falling off her seat and onto the floor, and then she stood up. She was shook up and hurt but had done her best to control the bus. Everett and Clay liked her.

The little old lady who had navigated the bus through the wreck on the icy intersection was tough. At 95 pounds with a smoker's cough, she was not only tough enough to hang onto the steering wheel with a broken wrist while dangling upside down, she was also tough enough to stand up to any trouble causers regardless of their sex or size.

There was a big Navajo boy from a rough family who lived out west of Spiderweb six or eight miles who started causing trouble one day as the little old white-haired lady drove the bus north over the summit in the late afternoon. The boy was in high school and four or five years older than Everett and Clay. He was getting rowdy, and finally the little old lady had enough. She pulled the bus over and stopped on the side of the road near the foot of the summit on the north side near the old Sacred Mountain Trading Post. She unbuckled her seat belt and turned around and faced the big kid who was twice as big as she was. "Get off the bus!" she said. The boy looked at her in surprise. She opened the bus door and repeated the order, "Get off the bus!"

He realized that she was serious so he stood up and walked down the center aisle and stepped off the bus. The little old lady turned back around, buckled up, and put the big bus in gear. The boy had walked down the road 20 yards in front of the bus and stuck his thumb out in midair, and before the little old lady had pulled the bus out onto the highway, a Ford pickup full of Navajos stopped to give the truant a ride. The school bus pulled out and passed the pickup and hitchhiker, and then after going a mile down the road, the Ford pickup passed the bus with the big boy sitting in the back of the pickup waving and laughing at the bus as the Ford pickup drove by. He arrived in Gray Mountain 30 minutes before the school bus.

One day when the bus left the last stop in town, which was Coconino high school, and headed north toward Gray Mountain loaded with kids, electricity was in the air. There were two Navajo girls, both seniors in high school, who were wired with 440 volts and, no doubt, supercharged with some kind of narcotics. They were gunning for each other and everyone could tell from their wicked expressions that there was going to be warfare.

When the bus stopped at the Wapatki turnoff to let the Rodgers kids off something happened to set the two off, and as the bus pulled away from its routine stop, they went at it. The scene was brutal, a nail-scratching, eye-poking, hair-pulling cat fight with no holds barred and death as a destination. The little old woman stopped the bus, put it in park, walked calmly down the aisle and reached the fight as one of the girls sat straddle over the top of the other. With a death grip in the other girl's long black hair, she was slamming the head of her opponent, who was pinned underneath her, onto the steel floor of the bus in an attempt to crack her skull open. Drugs were fueling her fury. The old lady with the hacking cough stepped up on the edge of the seats beside the girls so she straddled the one on top from above. She threaded her fingers through the girl's hair and braced her feet for leverage and jerked the witch loose by snapping her head back hard enough to addle her into submission; the girl on the bottom had had her head slammed against the floor enough to take the fight out of her. She separated the girls and then went back and sat down at the helm and calmly drove away.

On another afternoon during the trip going back home after school, two Navajo girls from Gray Mountain got into a fight on the bus. They were high-school-age girls and perhaps were a little high on drugs. The aggressor was a tiny little thing about five-foot high and 90 pounds soaking wet, and she was quite violent. She attacked another girl as the bus was traveling north of town on the way to Gray Mountain. The bus driver pulled the bus over and parked alongside of the highway near the Sunset Crater turnoff. The driver stood up and went back to the two fighting high-school girls and got the fight stopped. She told the aggressive girl to come forward and sit at the front of the bus. She refused to move, so the driver called the bus barn on her radio and told them to call the sheriff; and so as they waited for the sheriff to show up, the bus remained parked alongside of the highway.

Jamie Howell and Jean Ann had been in Flagstaff shopping and were on their way home. They drove up on the bus that was parked alongside of the highway so they stopped. Jean Ann got out and walked up the side of the bus and spoke to the driver through the closed door and inquired if everything was all right and asked if they could get the boys and Jamie's daughter off the bus. The driver said they were waiting for the sheriff, and until he got there, the door of the bus was going to remain closed, and no one was getting on or off.

Inside the bus, the Navajo girl, who was the toughest, told the bus driver to open the door and let her off, and she would walk home; but the driver said no, and so for 15 minutes or so they all just sat and waited, including Jean Ann, Jamie and Jamie's pre-school-aged girls.

When the deputy sheriff arrived he entered the bus and walked down the aisle and told the tough Navajo girl to follow him outside, and so they walked toward the door of the bus with the deputy leading. There was a decline outside the bus door covered by lots of cinders, which made the footing very precarious. About the time the deputy's feet hit the ground directly out of the bus door, the Navajo girl swung her backpack into his back knocking him down as she rushed past him trying to escape. The deputy's feet slipped out from under him, and he went down. A second later he sprang up, grabbed the girl, slammed her to the cinders and handcuffed her. It all happened in a few seconds. The deputy called Jean Ann that evening and asked her if she thought he had been too rough on the girl. She assured him that from what she had seen, he had done only what was necessary to handle the situation. For the kids, it was just another day on the Gray Mountain school bus.

For a while the Gray Mountain school bus was driven by a middle-aged man who wouldn't control the kids, especially the bigger boys who rode the bus all of the way to Gray Mountain where the bus turned around and started the return trip to Flagstaff 45 miles away. The bigger boys had the intuition to see and understand that the driver was not going to maintain any discipline, and the atmosphere went downhill fast. The big, tough boys ruled supreme and antagonized anyone and everyone who wouldn't bow down to them. The smaller younger kids were afraid.

Rather than confront the rowdiness, the bus driver tried to appease the bullies, and he told them he would play any kind of music over the PA system that they desired. Soon heavy metal, death metal, and every kind of evil garbage available was being played loud. The driver also told the bullies that they could decorate the bus however they wished. So, over a matter of several days, the windows of the school bus were blackened out by posters of AC DC and every other fringe and extreme band; the posters being glued and taped to the windows and walls, and subsequently no sunlight was allowed inside. The bus was a very large and long machine, and its interior became like a dark cave, its walls reverberating with loud death-metal music as it went down the highway.

Everett and Clay talked about it and talked about the thugs who congregated in the back of the cave doing any kind of activity they wanted. Everett was ten years old and Clay nine. Vic Howell's oldest daughter, Victoria, was also riding the bus, and she was seven years old. The boys convinced me that the bus was out of control.

One cold winter morning, I drove Everett, Clay, and little Victoria out to the bus. This was a job usually done by Jean Ann or Vic's wife, Jamie, but on this day I took the kids out to the bus stop which was a mile from the house. When the bus pulled up and stopped Everett, Clay and Victoria got out of my pickup and started walking toward the door of the bus, and I went with them. The door of the bus opened and the kids started up the steps one at a time, and I started to follow them.

The driver tried to shut the door before I could get in saying, "Hey, you can't get on here."

I raised my arm and shoved the folding door back open and stepped into the opening and started up the stairs with the driver repeating his order that I couldn't get on the bus. I wasn't interested in anything he had to say and turned my attention to the big bullies at the back of the bus. I looked down the aisle into the dark abyss, "The next time I hear about any of you tough guys pushin' these little kids around, I'm going to come to Gray Mountain and beat you until blood runs out your ears! Do you understand?" I turned to the driver who was as white as a ghost. "If you don't know how to keep order on this bus maybe I could bring you a cedar fence stay and use it on you to show you how!" I turned and walked down the steps and got back in my pickup.

Jamie Howell happened to go to town that very day and went to the school to pick up her daughter. Mr. Flick, the principle of the elementary school where the boys and Victoria attended, saw Jamie and asked if he could speak to her. He stepped to the side of all the bustle of kids leaving school and said to her, "Jamie, you need to tell your uncle Ed that he can't be getting on the bus and threatening the driver."

Jamie said okay, but she thought to herself, "Why don't you tell him yourself!"

The boys reported that the very afternoon after my visit with the bus driver all the posters had been taken down and rolled up and were placed in a five-gallon bucket by the driver's seat. He informed the kids they could take their posters home with them that day or they would be thrown away; and there was no more death-metal rock music blaring over the PA system ever again.

Soon after that a new bus driver was hired, the new one being an old white-haired man about 70 years old. He couldn't keep things totally under control, but for a while the bus was a little quieter. Everett and Clay were getting older and more able to take care of themselves. They looked out for each other as well as their cousin Vicki; filling up one seat with Clay by the window, Vicki in the middle and Everett by the aisle.

One morning the kids got on the bus, and Dwayne, one of the biggest boys, sat immediately behind the new driver, which was unusual because he usually sat farther back in the bus. When the bus got going down the highway, Dwayne took a pistol out of his coat and held it directly behind the back of the driver's chair where the driver couldn't see it. Everett was across the aisle and could see what was going on. Dwayne began pulling the trigger of the unloaded gun and dry firing it, which made a snapping sound loud enough for the bus driver to hear. "What's that? What are you doing?" the bus driver would say as he looked in the large rearview mirror directly above his head. He could see Dwayne looking back at him in the mirror and grinning. Then Dwayne would pull the trigger again. "What's that?" the driver would ask as Dwayne laughed at him. This went on all the way to town, but the driver never had the courage to get up and demand to see the pistol. He knew what that sound was.

When Everett got off the bus at the junior high, he slipped away so the other kids in that boy's family wouldn't know what he was doing, and he went to the principal's office and reported what he had seen. The principle got on the phone and called the high school, which was the last stop for the bus and where he would have to get off, but by the time the authorities were notified Dwayne and the pistol had disappeared.

That was the last day Dwayne ever rode the bus or attended school. Everett never knew what the story was behind Dwayne's actions. Maybe it was his graduation and departing party for himself. Why did he have a gun? Maybe he was going to collect a debt, or maybe someone was going to come calling for a debt he owed. Everett figured there were some bullets in his pocket and he was going to load the gun at some point, and his last ride on the bus was for torturing the bus driver.

There were three big boys in the clan of Navajos who lived west of Spiderweb that were older than my boys by four or five years. They were tough, and the other hooligans on the bus gave them plenty of space. These boys never bothered Everett or Clay or any of the other ranch kids.

There was some other hoodlums that lived halfway to town by a trading post called Sacred Mountain at the foot of the summit on the north side. These thugs were always in cahoots with several other bullies who lived at Gray Mountain. The chief among these Gray Mountain boys was named Glenn. Glenn and the Sacred Mountain boys were real trouble causers and liked to push people around.

Once in awhile, Glenn would get off the bus and stay with his hoodlum friends at Sacred Mountain rather than go all the way home to Gray Mountain.

One morning the boys got on the bus, and several miles down the road Wayne and Dwayne and their little brother from the clan west of Spiderweb got on. Trouble had been brewing between Wayne, Dwayne, their other brother and Glenn and the Sacred Mountain clan. Glenn had gotten off at Sacred Mountain the night before. Wayne, Dwayne, and their brother, who looked kind of bruised about the head and face, sat up near the front when they got on, which was out of character. Ten miles farther the bus pulled up to Sacred Mountain, and there stood Glenn and the Sacred Mountain clan; and they started walking toward the door of the bus. Wayne, Dwayne, and their little brother stood up, walked to the door of the bus, and quickly stepped outside meeting Glenn and the Sacred Mountain boys who were totally taken by surprise. Wayne, Dwayne, and the brother immediately went to pulverizing them. It was bloody. The bus driver shut the door, put it in gear and drove off leaving the six of them rolling around in the cinders on the edge of the highway. The next day Glenn and the Sacred Mountain boys had lots of knots on their heads.

There was a naïve high school boy named Sean living at Spiderweb who chose to go to the back of the bus and hang out with Glenn and listen to Glenn's death-metal music and come under the influence of Glenn's dark agenda. This agenda would soon reap dire consequences.

Chapter Thirteen

February 12, 1988 was a tragic day on the Babbitt Ranch. The story began several years before when Bill hired a man by the name of Don to fill the position of maintenance man on the Babbitt Ranch. Everyone on the ranch called the person with that job the waterman rather than maintenance man because his main duty was to keep water running down the many miles of pipelines that scattered water all over the hundreds of square miles of the ranch and filled many dozens of water troughs and huge water storages with water that numbered in the millions of gallons. It was an important job but sort of a thankless one. The waterman usually worked alone at a task that was tedious and devoid of any glory but was still very important. Usually the waterman did not mix much with the cowboy crew and did not share in the camaraderie enjoyed by those men.

Don was a good man and well qualified, and Bill and Vic Howell liked him. He had a very nice wife named Ronda who got a job in town and drove back and forth every day to fulfill that obligation. Don and Ronda had an adopted son by the name of Sean who was going to turn 18 later on that year, but in February he was 17 and attended high school in Flagstaff and rode the Gray Mountain school bus. Don and Ronda got custody and adopted Sean when he was 12 years old.

Sean was a problem child who had been passed around from one foster home to another, each one ending in a bigger disaster than the one before until at the age of 12 when Don and Ronda adopted him. He was a very mixed up, troubled and confused child.

Don had been a Marine and had taken his Marine discipline serious, and thinking that this type of mentoring is what young Sean needed, he was trying to instill some form of military boot-camp style parenting into he and Sean's relationship, but it didn't seem to be working. Sean was uncoordinated, always late, and the more his drill sergeant stepfather screamed at him, the less interested in work he seemed to be. He was completely unmotivated, and Don, the father, became increasingly more frustrated and at his wit's end trying to get Sean excited about life. Sean did not do well in school, did not do well at any job or task set before him. He began hanging around the losers, the dropouts, the kids doing drugs. There were several bully boys who lived in Gray Mountain and rode the Gray Mountain school bus who drank, smoked dope, listened to hard rock music, and pushed their Satan-worshipping agenda into Sean's weak mind as he rode the bus. He was on a downhill slide.

On the afternoon of February 12, Sean did not ride the school bus out to the ranch but, instead, walked from Coconino high school down to where his adopted mother, Ronda, worked in the vicinity of Santa Fe Avenue and Fourth Street and caught a ride with her out to Spiderweb where they arrived about 5:45 in the afternoon. In later testimony, Ronda said that she and Sean went into their house where Don already was, and they had a few minutes of family time just visiting and making small talk. She said that everything seemed to be fine and there were no ill words spoken by anyone.

After their short and pleasant visit, Don told Sean that he needed to go outside and do his evening chores, which amounted to feeding and caring for several dogs that were family pets. Ronda got up and went into the kitchen and started preparing supper. While outside, Sean fed the dogs and then opened the door to Don's pickup and got a Ruger .22 caliber pistol that was laying in the seat and proceeded back to the house. He opened the door that went straight into the kitchen where Ronda had her back to the door. She was facing the stove cooking supper, and when she heard the door open she turned and saw Sean pointing the pistol at her, and he said, "Well, I guess I've gone crazy," and then he shot her in the abdomen. Don immediately jumped up off of the chair he was sitting in, about ten feet away and to Sean's right, and ran forward toward Sean. Sean shot him one time, and he went down.

With Don and Ronda laying on the floor, Sean thought he had been successful in killing them both; and so he stepped back outside. Ronda managed to stand up and call Vic Howell's house that was several hundred yards away to the west. Ronda was, understandably, pretty hysterical and was screaming into the phone. Jamie, Vic's wife, answered and heard Ronda saying, "Don's dead! Don's dead!"

Vic Howell immediately jumped up and got in his pickup and drove to Bill Howell's house that was several hundred yards farther west. Vic and Jamie both thought that Don had suffered a heart attack and was either dead or dying. Ronda was not an emotional or unstable person but, instead, was cool and level headed most of the time, so because of the hysterical way she sounded on the phone, Vic and Jamie knew something was seriously wrong. Vic honked his horn and aroused Bill, who like everyone else on the ranch was relaxing after a full day's work. It was right about sundown.

Jamie jumped in her car, a Pontiac Bonneville, and raced down to where Jean Ann and I lived next to the bunkhouse and horse corral. I was inside the house with Everett who was 10 years old. Jean Ann and Clay were in Flagstaff. Clay had stayed after school to attend a friend's birthday party. Jean Ann had driven into town to pick him up, and they weren't expected to be back out to the ranch for an hour or two.

Jamie came driving rather wildly up to the house honking the car horn. I stepped outside as she stepped out of her car and hollered at me, "Come on, Don's had a heart attack and he's dying!" So I looked at Everett, who had stepped outside onto the doorstep next to me, and I told him to stay there, and I got in the Pontiac car. Jamie had punched the accelerator, and my right foot was dragging the ground as she raced off toward Don and Ronda's house, which was about 600 yards away to the north and on the other side of the old shop building. *Man, this girl can drive,* I thought to myself as we raced on toward Don's house.

As we got within view of the house we could see Ronda running around several parked vehicles, and Sean was chasing her. It was almost as if they were playing tag or some children's game. Vic and Bill had driven up only seconds before Jamie and I did, and Ronda came running to the driver's door of Vic's pickup; and Sean gave up the chase and started running north toward the Spiderweb horse pasture. After listening to Ronda for a few seconds, Vic jumped out of his pickup and screamed at Jamie, "Get outa here! Sean's got a gun! Get outa here!" I was half out of Jamie's car and was trying to stand up, when, for the second time in two minutes,

Jamie stepped on the throttle and threw gravel for 50 yards and almost drug my right leg off again. *Man, this girl can drive!*

Don and Ronda and Sean had been living in a trailer house situated so the doors opened out toward the north. The door located farthest to the west opened up into the kitchen/living-room area of the house that was 12-foot wide and 60-foot long. There was a hallway that ran down the north side of the house and toward the east end of the house there was another door opening out toward the north. When Sean entered the door into the kitchen and living room, he shot both of his adopted parents, and thinking them both dead, he exited that same door that led onto a covered porch that ran along the north side of the house. After reaching the outside he could hear Ronda talking loudly into the telephone, talking to Jamie, and so Sean entered the house a second time and began assaulting Ronda, shooting her at least once more.

Ronda broke and ran down the hallway and went out the door on the east end of the house and continued running out and around the several parked cars with Sean chasing her holding his pistol in his hand, but he had run out of bullets. This is how the scene looked when Vic and Bill drove up to the house with Jamie and me arriving only several seconds later.

As Jamie and I sped away, leaving the scene, I saw Sean running north in the direction of the horse pasture and many miles of open country. Vic and Bill successfully loaded Ronda into Vic's pickup and drove off to Bill's house where they got Ronda inside and laid down on a bed; and Bill's wife, Linda, and Jamie began ministering to her as best they could. She had several bleeding bullet wounds in her abdomen. The county sheriff's office was notified as well as an ambulance.

Jamie had quickly dropped me off at my house, and I got out of her car and told the several men in the bunkhouse as well as the old woman cook, Betty Rodgers, what all had taken place. Betty came over to our house, and she and I and Everett sat for a few moments talking things over. We all wondered what would transpire next. We didn't know if Sean had more ammunition or, perhaps, several firearms or if he was wanting to go on an extended attack on everyone else at the ranch. We wondered what in the world had caused him to act like he had acted. There were a lot of pieces of the puzzle we didn't know.

In about five minutes Bill Howell drove up in front of my house in his pickup. I believe he honked the horn, getting everyone's attention. I quickly walked out to his truck, and he told me, "Ronda says Don got shot and is down, but she isn't sure if he's dead or not. We need to go over there and see what has happened. If he's alive we need to help him." By the time he said this, the men in the bunkhouse came out and were standing by. These men were Charles Kent, Tad Dent, Brian Thomas, and Matt Thomas. Bill gave those four men orders to get in two pickups, two men in each truck, and we were to surround Don and Ronda's house and shine the headlights toward the house, one truck pointing toward the house from the west, one truck from the south; and then he told me to get in with him, and we would point Bill's truck toward the doors from the north. He and I would have the best view of the house's doors and front porch. The sun had now been down long enough that it had gotten dark. Before getting into Bill's pickup, I ran back in my house and told Betty Rodgers and Everett to stay inside and lock the doors.

So us six cowboys drove over and surrounded the house, flooding it with pickup headlights. We didn't know where Sean was, or if he was still armed, or what he might be armed with. We thought that maybe he had gone back inside the house and was barricaded in, waiting to shoot anyone who entered.

Bill and I sat for several minutes staring at the porch and the doors of the house, both of which were closed. Finally, Bill told me, "I'm going in there!"

"Okay," I said.

He jumped out of the driver's side of his truck and moved quickly toward the nearest door, the one on the east end of the house. I drug a lever-action Winchester out of a scabbard that Bill always kept under his pickup seat. I jacked a cartridge into the chamber and stepped out and stood by the passenger side of the pickup. *Where was Sean?* The last time I had seen him he was running north about 15 or 20 yards from where I was now standing. But where was he now?

Bill arrived at the door of the house and hollered very loud, "Sean, Sean! Are you in there?" Nothing but silence came out of the house. Bill banged real hard on the door with his fist and repeated, "Sean, are you in there?" Bill kicked the door open with his right foot, and from where I stood, about 20 yards away, I could see a light in the living room down a dark hallway. I could see a body laying on the floor. Bill hollered Sean's name again and heard nothing in return, and then suddenly he broke into a run and went down the hallway and into the living room. Within a few short seconds, he came running back outside and got back in his truck. "He's dead," Bill said. "Got a bullet hole right in the middle of his forehead!"

There was now no reason to have the house surrounded and flooded with lights, so all six of us cowboys went home. Charles, Brian, Tad and Matt to the bunkhouse; while Bill, after returning me to my home, returned to his house where Linda and Jamie were trying to keep Ronda alive as she lay on a bed waiting for an ambulance to arrive. Vic had been stationed there at Bill's house to be on guard as they all watched for the ambulance.

The last time anybody had seen Sean was when Vic and Bill, with Ronda loaded in Vic's pickup; and Jamie and I in her car, had all sped away from the crime scene watching Sean run north from the parked cars outside of Don and Ronda's house. But none of us really knew where or how far he had gone. Later we learned that he had crawled through a barbed-wire fence just a few feet north of the house and ran a short distance out into the horse pasture, and then as evening grew darker, he turned to the west and walked three or four hundred yards and turned back south and crawled back through that same fence; and from there, it was only several hundred yards south to Bill's house. Sean later said that he looked through a bedroom window and watched Linda as she bathed Ronda's bullet wounds. He would have been doing this at about the same time that Bill and I and the other four cowboys had our truck lights shining on the house that was the scene of the crime.

After watching his wounded mother through the window for a moment, Sean started walking south. He had run out of bullets and dropped his gun. He obviously had no plan B, and he was now wandering aimlessly.

It took a half hour or more for any law enforcement personnel to show up, but when they did there were several of them. Deputy sheriffs came and talked to me as I sat in our house with Everett and Betty Rodgers. They asked what I knew, which amounted to very little. They proceeded to walk around the barns and corrals in pairs holding flashlights, searching for a 17-year-old boy that everyone mistakenly believed was armed and dangerous. After a while a helicopter flew over shining a big spotlight all over in an attempt to secure the area and make it safe for the ambulance that would not drive into the ranch until it had been declared safe. That took a while, probably an hour.

About the time the ambulance was allowed to drive in and pick Ronda up, a big police car drove up to my house, and a small man with a rusty-colored moustache and a cowboy hat got out and knocked on our door. It was Coconino County Sheriff Joe Richards. This was the first time I met Sheriff Richards, although we would eventually become close friends. I invited him inside, and he asked me where I thought Sean was.

"I bet he's out at the dump in the middle of the horse pasture. There is several old car bodies out there, and he could get in one of those old car bodies and hide. The last time anybody saw him he was headed that direction, north. It's about a mile out there," I told the sheriff.

"Could you take me out there?" the sheriff asked.

"Sure," I said

"I bet he's up on Black Rock Hill, Dad. Hiding in those caves," Everett spoke up.

"Oh, no he isn't. He's out at the dump," I told Everett in a tone of voice that suggested he should mind his manners and be quiet.

"Where's Black Rock Hill?" the sheriff asked.

"Oh, it's that big hill that is between here and the highway," I told the sheriff, and I pointed in a south-westerly direction. Black Rock Hill was straight south of Bill's house about a half mile.

"I bet he's up on Black Rock Hill, Dad," Everett said.

"No, he's out at the dump," I said then turned toward Betty Rodgers, "You and Everett stay here and keep the doors locked, and I'll take the sheriff to the dump."

So Sheriff Joe and I stepped out of the house leaving Betty holding on to Everett with both her hands on his shoulders as he stood in front of her looking at me wondering why I wouldn't listen to him. The sheriff and I drove north through the center of the horse pasture and arrived at the dump, which was no more than a spot where refuse had been deposited over the side of a malpai rimrock where it had accumulated in heaps over numerous decades. Scattered around were a half dozen old car bodies, mostly old, worn-out ranch pickups and maybe a car or two abandoned by some ranch employee. It was my idea that Sean would be hiding in one of these abandoned cars; that idea being based upon the fact that the last time anyone had seen Sean he had been headed toward that spot, even though his last point of being viewed was a mile to the south.

Sheriff Joe got out of his big police car with a flashlight in his hand and told me he was going to look around. He was not wearing a firearm of any kind, but there was a snub-nosed .38 special Smith and Wesson pistol laying in the seat of the car. I had no interest in getting out of the car, and he didn't give me the impression that he needed or wanted any assistance. He looked at me and said, "Use that pistol if you need to," and with that he walked out into the night. It was very dark. After walking around and shining his light in every car or other hiding spot, he returned empty-handed and, we drove back to my house. By that time every building and corral at the ranch had been thoroughly searched by several deputies who had been wandering around in pairs flashing their flashlights into every nook and cranny of the place.

Sheriff Joe left and went back to Flagstaff, but there was still several deputies wandering around. The ambulance came and hauled Ronda into the Flagstaff hospital, and we all wondered where Sean was. I was worried about Jean Ann and Clay, thinking that they should have returned; and so I went and talked to Matt Thomas who was a single boy living in the bunkhouse. I asked him if he would drive out to the highway, which was a mile away to the west, and see if he could see any sign of her. So Matt left, agreeing to do that errand for me, while I sat in the house with Betty Rodgers and Everett. By now several hours had passed since the shooting, and Jean Ann and Clay were overdue by at least an hour. Another hour passed, and Matt Thomas nor Jean Ann and Clay had appeared. What in the world could have possibly happened to Matt I wondered to myself, I only wanted him to drive one mile and return. So I left Betty and Everett in the house, instructing them to keep the doors locked; and I got in a pickup and drove out to the highway, and there I found a highway patrolman who had established a roadblock not letting anyone into the ranch. Jean Ann, Clay, and Matt Thomas sat in the front seat of our pickup with the heater running looking like refugees wanting to return to their homeland. I drove up, and Jean Ann rolled down her pickup window and said, "They won't let us go down there; they say it is too dangerous."

"Oh bologna," I said. "Matt, get back in your pickup and we're all going home." I told Jean Ann to put it in drive and follow me, and she, Matt and I all drove though the Arizona Department of Public Safety roadblock and went home, leaving the highway patrolman standing and staring at us in disbelief.

The rest of the night passed uneventful except the fact that Everett and Clay, now that he was home, both told me that Sean was up on Black Rock Hill hiding in one of the caves in the side of the mountain. I ignored them thinking that Sean had, without a doubt, continued north toward Gray Mountain and the Navajo Indian reservation.

Early the next morning, after graining the horses, Matt Thomas and I loaded a ton of cow feed in 50-pound sacks onto the back of a flatbed and drove west toward the highway with the intention of going up into 89 Pasture and feeding the thousand head of yearlings that I was taking care of. We drove past Don's house, Vic's house, and then Bill's house and saw several Coconino County sheriffs' vehicles parked and several deputies walking around looking for clues. When we got a quarter mile past Bill's house, we looked down the road and saw Sean walking toward us in the direction of all the houses we had just passed. I stopped the truck and turned around and drove over to where the law enforcement personnel were and told them the man they were looking for was walking their direction about a quarter mile away. They all jumped in several marked cars and sped off toward the highway and the 17-year-old boy who had committed murder. Matt and I followed behind and watched the deputies stop several hundred yards away from Sean, jump out of their vehicles, and draw their guns. They also released two very large German Shepard police dogs that ran very aggressively at Sean who seemed frozen. The deputies screamed, "Get down! Get down! Lay on your stomach and put your hands over your head or these dogs are going to kill you! Get down! Get down!"

Sean lay face down in the middle of the Spiderweb road with two very big dogs standing over him with eyes of fire staring though him like lasers. The deputies ran up and put Sean's hands behind his back and handcuffed him.

When all of this had been accomplished, Matt and I drove by and passed within six feet of where Sean stood surrounded by policemen who were frisking him. He was totally defeated. Gone was the arrogant craziness and fantasies of the afternoon before that had whispered into his ear and convinced him to get a gun and shoot someone. He was wearing a canvas Carhardt coat that was worn out, and it was about 10 degrees outside. He looked like he was about to freeze to death. There was no fight left in him. It was one of the saddest things I've ever seen. He had spent the cold night in a cave on Black Rock Hill.

Chapter Fourteen

Every ranch in the American West has a certain flavor that has attributes poured into it by the man who runs it, the cowboys who work there, and the terrain the cows graze on, and even the cattle themselves. Babbitts, at least when I worked there, was a roping outfit. Bill Howell was a champion roper who had successfully competed as a team-tyer, dally team roper, calf roper, and single steer roper. He was also an excellent horseman who could train a rope horse as good as anyone going down the road. I saw Bill rope several thousand steers by the head, and I have no memory of him ever missing a head loop. I suppose he did, but I honestly cannot remember it happening, and when it did it was a rarity. When you worked cattle on the Babbitt Ranch you were going to get a chance to use your rope, and, heaven forbid, if you didn't like to rope you didn't fit in; that's just the way it was.

Bill was a perfectionist. When we branded in the spring we usually had a ten-man crew and sometimes eleven or twelve. Bill liked to have two men heeling and dragging calves to the fire and two sets of flankers, one on each side of the fire. I have seen him many times standing and leaning against a tall propane bottle and checking his watch, timing the branding of calves as it took place. If his ten-man crew wasn't branding two calves a minute he wasn't happy. That feat cannot be accomplished with a crew of amateurs. Sometimes we would brand 2.2 calves a minute and even occasionally 2.5 calves a minute, which would really make him happy. If we branded outside, without a corral, with one man roping the calves around the neck and dragging them to one set of flankers, he considered branding one calf a minute acceptable. On several occasions we branded 150 calves in just under three hours outside with no corral, and he was happy about that. If the herd became scattered because of a roper getting too wild in the herd, he was mad, but that never happened because someone who was too wild never got to rope. He made sure of it.

In 1981 the Babbitt cowboy crew, myself included, built a roping arena at Spiderweb. Everett was four years old and Clay was three. The Babbitt remuda was becoming a bunch of nice horses and evolving out of a bunch of meat-headed buckers. Bill bought the first good Driftwood stud in January of 1975 and started the now famous Driftwood breeding program. The ranch received the AQHA Remuda of the Year award in 2005 which was a result of this decision of Bill's to try and raise some good horses.

When I first went to work for the outfit in 1974 the Babbitt Ranch had lots of horses, but they were noted for being double tough, big, hard-mouthed jugheads; and lots of them bucked. A lot of cowboys didn't want to work there because the horses had such a bad reputation. So Everett and Clay grew up on a roping outfit with a roping arena close by and plenty of good ropers to watch. It was only natural that they would become ropers.

Everett came out of the womb motivated. He worked at being a top hand and wore out many ropes just throwing them at a sawhorse dummy. Clay came into the world laid back, sort of casual, not caring about championships or practicing to get there. While Everett fretted over throwing the perfect loop at a practice dummy, Clay would be stalking a cottontail rabbit with a BB gun or trying to catch a bird that was flying around the rafters in the barn.

Everett set out to be a world champion heeler at an early age, and in those days Clay Obrien Cooper was considered the best that had ever lived. The conversation at the practice arena was always about Jake Barnes and Clay Obrien Cooper, the living legends who were setting new records that none of us thought could be broken by anyone—ever!

When the kids were about eight or nine years old, I built a heeling dummy sled out of scraps of lumber I found laying around the ranch. Basically it was a sawhorse on runners, similar to a sleigh, that I could pull with a horse while Everett and Clay followed behind on Charro and Beaver, two young Driftwood horses that were in their strings; and they would take turns throwing heel traps in front of the back legs of the sled as I drug it along. On many days I would have Charro and Beaver saddled when the boys got home from school after a ride on the school-bus-from-Hell, and I would drag the heeling sled out to the highway and back as they rode along taking practice heel shots at the sled. It is a mile from the barn to the highway. You can make many a practice throw in two miles.

After a few months of that, the wooden heeling dummy exploded into a pile of splinters, the result of a hard and violent life; so I built a new one out of pipe. I've built several pipe heeling dummies since then. Everett and Clay both learned how to handle their slack, dally, and control their horses roping one of these heeling sleds. They both have heeled a sled at least a million times.

Everett, being focused and singular in his devotion to becoming a world champion cowboy, got more attention from everyone, including me, at least pertaining to anything related to the craft of roping. Clay was just a happy-go-lucky kid, just as content with a BB gun as a horse and rope. Everett wanted to be a heeler like Clay Obrien Cooper, so we were all trying to help him be a heeler. Clay had to follow along and get the crumbs of everyone's attention. If Everett was going to heel down at the arena, Clay had to heel also. I gave no thought to it, instead; it just seemed a matter of course.

One afternoon we had been down at the arena roping, and both Everett and Clay were doing nothing but heeling behind one of the several headers who were there, which was probably Bill or Vic Howell, or myself. Clay didn't seem to be having much fun.

That evening, after we got into the house, I asked Clay how he was doing; and he indicated that he was not feeling good. I sat down in my easy chair, and he got up in my lap. I asked him to tell me what was going on. "I don't feel good about myself," he said. He was about nine years old.

"Why not?" I asked. This kind of statement coming out of Clay was most unusual, and it got my attention.

"I have to heel all the time. I don't want to be a heeler. I want to be a header," he said.

I realized that no one had ever asked him what he wanted to do, but, instead, I just saddled our horses and he was told to get in the heeling box. I suddenly felt ashamed of myself. I hadn't been paying attention. I didn't have a clue what was on my kid's mind. I wanted to cry. "Okay, you can start heading. We will go to working on it," I said. And so the course of history was changed and a very good header started his career.

Chapter Fourteen

Clay was comical. He saw things differently than the rest of us. The first time Clay entered the twelve-and-under team roping at the Cowpunchers Reunion he was probably about eight years old. I was heading for him, and we drew a really rotten steer that ran like a deer. There was absolutely nothing I could do with the steer when I roped him. He was wild and would run up the rope. Clay was having a very hard time keeping up as I got the steer on a short rope and tried to slow him down, but the steer would not cooperate. Art Savoini was flagging. He rode around holding the flag up for what seemed like 10 minutes as Clay tried to maneuver into position. Both my kids had been schooled about never roping out of position. "Don't ever take a bad shot!"

We circled, and the steer jumped and ran and did somersaults, and Clay lumbered along trying to catch up. He was riding a horse named Waco who was a real plug. Soon after this Clay graduated up to a much better horse, but there we were burning up the clock as Clay and Waco strove for position. Finally the steer sulled up and stopped and just stood still. I thought that finally Clay will be able to throw a trap in front of the steer's hind legs, but he rode up into perfect position and swung his rope backwards and whipped the steer down the center of his back. "Hee haw you sorry son of a gun! Hee haw", Clay said. And the steer took off running. Art Savoini almost fell off of his horse laughing. Eventually we had to content ourselves with a no-time.

Several years went by, and Clay got to heading real good, especially for a little kid of 12 of so. We went to all of the Arizona Cowpuncher Reunion Association ropings as I was very involved and a member of the board of directors for many years. In those days Billy Hamilton would come to all of those ropings and was always in the winner's circle. Billy Hamilton had been a world champion team roper, and even when an old man he roped very well. Billy would always be sitting on his horse down at the end of the arena hiding behind his sun glasses and very quiet, but he wasn't missing anything. He was a thinker. He watched Clay rope a few steers and decided he wanted to rope with him in some of the low-numbered handicap ropings. As Everett rode by him, Billy hollered at Everett and said, "Hey, I'd like to rope with your little brother." Everett certainly knew who Billy Hamilton was, and he almost started hyperventilating when the famous cowboy spoke to him.

"Yeah, sure," Everett said nervously. "I'll go talk to him." Everett quickly rode off like an angel messenger from God so he could give Clay the wonderful, almost unbelievable, news. He found Clay out in the parking lot playing with some kids. "Clay," Everett almost screamed. "Billy Hamilton wants to rope with you!"

"Who's Billy Hamilton?" Clay said as he casually looked up from some game he was playing in the dirt. He didn't care if Billy Hamilton wanted to rope or not.

Another time we were in Kingman at a big jackpot, and I was trying to find my kids some good partners for a roping. Brad Smith was sitting out in the middle of the arena with several of his friends. Brad had just won the world, and he thought he was something. He had just got out of the beauty shop, and his hair was nice and wavy, and his shirt was starched. I told Clay to go ask Brad to rope, and I pointed at him. "Go ask that guy over there, his name is Brad Smith. He's a world champion." Clay was afoot and he walked out to where Brad and several of his friends were sitting on their horses visiting. Tanner Bryson was next to Brad. Clay, not knowing for sure which one was Brad, walked up to Tanner and said, "Hey, Brad, you want to rope?" Brad started jumping up and down in his saddle trying to get Clay's attention. After all he was the one with the expensive hairdo and gold buckle. Didn't this dumb kid from the sticks know who he was? Tanner played it perfect, acting like he was Brad and Brad was just some movie-star-looking guy who needed to go pee really bad. "No, kid, I'm full up." And with that Clay walked away thinking that the man with the hairdo was sure a funny acting dude.

Everett and Clay grew up mounted on good Babbitt ranch horses. The most noteworthy being Beaver, Waddie, Charro, Hot Shot; and later Kid and Scooter, both of whom eventually belonged to them, Clay owning Kid and Everett owning Scooter. Kid was a streaked-face red sorrel out of the first Driftwood stud the outfit owned that was commonly known as Speedy. Kid stood 15 hands and weighed 1200 pounds and was smooth as glass. He was a pro-level heading horse. Scooter was a little bay horse out of a Driftwood mare and a stud called Joe Burge. I can't recall his pedigree. Vic Howell raised Scooter but traded him to me because he had crippled front feet, but I could shoe him and keep him as sound as any horse around. He stood, probably, 14.3 and weighed 1050. He was smooth and was a pro-level heeling or calf horse.

Growing up, both boys were around men who headed steers really well, and so Clay received lots of schooling as a header. But Everett wanted to be a heeler, and the schooling in the heeling department was deficient. Harvey Howell, Bill's younger brother and therefore my brother-in-law, as well as a close friend, was one of the best arena heelers in Arizona, at least in the 1970s. But Harvey did not have a gift for teaching and explaining the basics of heeling. Bill Howell was also a good heeler but couldn't explain how to do it. Tim and Tom Howell, the younger two of Bill's sons both grew up to be good heelers. But the heeling instructions were very poor, mainly due to the lack of communication. Tim Howell promoted a heeling school with Billy Hamilton's son Craig coming to the ranch and acting as an instructor. That was a waste of time, at least in my opinion. Everett was his main student, and he didn't tell Everett anything except a bunch of mind-over-matter religious mental exercises and told him to keep a journal to record his deepest inner thoughts. Everett really got to roping good after he left home and worked for and lived with John McKenzie in Phoenix.

Somebody had a movie camera and would take movies of our practice sessions and later we would go up to Bill's house and watch the movies. The main thing I remember was everyone, after watching the films, would tell Everett, "See, Everett, you're dropping your shoulder. You need to quit dropping your shoulder." Everett was dropping his shoulder because the steers' heels were right beside his right stirrup. If he would have let the steer get out in front of him, which would have made him reach out and rope him, he wouldn't have dropped his shoulder. But somehow that was all lost in the communication process.

The Howells, every one of them, could heel and drag calves to the fire as good as anyone who ever lived. They were masters at it. But they couldn't tell a kid how to heel in the arena. If you watched them and studied what they did, you could pick stuff up by watching their example; but they weren't teachers. But Craig Hamilton was a top-level pro roper and he was no teacher either.

When Everett and Clay were ready to crack out into the minor leagues of competition, I was presented with the problem of how to get there. The ranch was good about hauling our horses to the cowboy reunion, but if we wanted to go compete somewhere else we needed our own rig.

When Everett was nine and Clay eight years old, we owned a big Chevy Suburban. It was four or five years old and a gas hog. I was on the Arizona Cowpunchers board of directors, and after a meeting at Bob and Darlene Burris' house in Chino Valley, I drove down to a car dealer in Prescott and traded for a long-wheel-base, half ton Ford pickup. We soon bought one of those little sleeper shells that fit behind the rear window. Clay loved to get up in there and sleep while the rest of us sat on the bench seat in front. This pickup was four or five years old and had about 60,000 miles on it when I bought it. I was making $650 a month cowboy wages, and Jean Ann wasn't cooking any longer, so we were living on my income alone

Now, we needed a trailer so I found one for sale in the Flagstaff newspaper and went and looked at it. It was a homemade affair with one axle and a big letdown ramp in the rear so horses could walk up it to get in. Horses hate ramps. This new trailer was extra wide with plywood sides and a big hood over the hay manger in front, but from the manger to the back, it was open topped. It was, without a doubt, the ugliest trailer ever built. I bought it. The boys named it Turkey Turd. Actually Turkey Turd was a compliment to what it really deserved.

Next I traded the Ford for a crew-cab, one-ton Dodge with a gas engine and manual transmission. It was four or five years old and had 60,000 miles on it when I bought it after a Cowpunchers Reunion board of directors meeting. I seemed to be setting a pattern. I traded Turkey Turd (after several years) for a 20-year-old Hale bumper-pull 5X16 with a metal top and wooden slats on the sides. It was quite rusty so Everett and Clay and their cousin Todd put a coat of white paint on it with brushes. The boys were now 13 and 14 years old, and I was making $800 a month.

I often worried and stressed over not being able to provide better stuff for my kids, and I wondered if the embarrassment of driving into the rodeo ground pulling Turkey Turd or an old rusty Hale stock trailer and parking next to a brand new Prevost bus that was 50-foot long and hooked to an aluminum trailer that cost more than a decade's worth of cowboy wages, caused irreparable damage to my two boys' mental health. At least we were always mounted on good horses. Our trucks and trailers were old and worn out but our horses were the best. And we had trained all of them ourselves.

When I bought the Dodge crew cab I thought I was really stylin'. It was a 1985 model and when new it would have been considered a pretty nice rig, but it was at least four years old when I got it.

I remember one roping we went to was the first USTRC roping ever held in the state of Arizona. It was in 1989 at West World in

Scottsdale. There was a huge blizzard that weekend and during that storm there was a pileup on I-40 on the east side of Flagstaff with over a hundred cars involved. It was caused by people, including semi-trucks, driving way too fast on ice with zero visibility. Several people were killed. On the way home we had to unload our horses, Kid and Charro; and Everett led them up the summit north of Flagstaff because the Dodge couldn't get enough traction to pull the Hale four-horse and two horses up the hill. The Dodge was a two-wheel drive.

This Dodge would get vapor locked because it was so long the fuel pump couldn't push enough fuel from the tank to the engine when the weather was hot in the summertime, so we added an extra pump into the line. Next the transmission blew up and then the engine had to be rebuilt. But by this time I had taken the wagon boss job at the Diamond A Ranch at Seligman, and I was making big money, $1500 a month. We bought a half-ton Dodge that was only a couple years old and a new single-axle, two-horse trailer. It was even made in a factory. We would show up in it instead of Turkey Turd, and it was less embarrassing when we had to park next to the $500,000 buses at the high school rodeos. We had arrived! Ha! In a used pickup, but the paint was good.

Chapter Fifteen

On October 26, 1991, Jean Ann and Everett went down to Prescott to attend the wedding of Justin Salcito and Dawn McFarland. Dawn's parents were old friends of ours, and she and Everett were pals. Her husband-to-be was a cowboy whom we didn't know but had heard good things about, and we also knew that he rode saddle broncs at the pro rodeos. The fall roundup was going full blast at Babbitts, and I was busy, so Clay and I stayed home at the ranch.

When Jean Ann and Everett got home from the wedding they had lots of good stories to tell about all the people they had seen and visited with. They had great fun telling Clay and me all the big news. Everett said to me, "They are having a bronc riding school this winter down in Phoenix, and I want to go."

"Who's having a bronc riding school?" I replied with as much stern negativity as I could muster.

"Well, Lyle Sankey is putting it on, and he's teaching bareback riding. Cody Custer is teaching the bull riding, and Cody's brother Jim Bob is teaching saddle bronc riding. Justin is going to help him. All of them guys were at the wedding, and they want me to go. Steve Dollarhide will probably be there. I want to go."

"I thought you wanted to be a world champion team roper?"

"I do, but I wanna ride broncs, too." Everett said.

"Yeah sure, you want to ride broncs. What will happen is you will go down there and get bucked off three or four times, and you will lose interest. And besides that, we can't afford it. If you want to ride broncs, I'll teach you to ride broncs, but we're not going to a bronc riding school." I had already made up my mind: Everett wasn't going to any bronc riding school.

Everett left the room looking dejected, and Jean Ann gave me a look of disgust. I mumbled to myself, "I can teach the kid to ride broncs if that's what he wants! I've ridden lots of bucking horses." I was probably talking to myself out loud. Everyone had left the room.

Later that night when we were alone, Jean Ann let me have it, "What do you mean, we can't afford it? You have the worst poverty mentality I've ever heard of! Let the kid go to the bronc riding school. We always seem to have the money for you to enter a roping when you want to. Let him go to the school!"

The topic of rodeo schools, bronc riding in particular, stayed around the house for a few days. I held to my stand that we couldn't afford it. Besides within several buck-offs, the bronc riding would be out of Everett's system, and I wasn't going to pay for a short-term pipe dream. Not on Babbitt Ranches cowboy wages. Jean Ann kept telling me to let the kid go. Everett's interest didn't seem to wane, even after a few days. Finally, I relented, "Okay, we will take you to the Sanky rough-stock school. But you better not break your leg. We can't afford that!" That was the story of my life; I was always worried about money.

We went to the Lyle Sankey rough-stock school, which was a three-day event held at the Estrella Park arena near Goodyear, Arizona, There must have been 40 or 50 students spread out between the bull riding, bareback riding and saddle bronc riding. There were several ranch kids there that we knew including Luke Leist, Pete Criner, and Brooks Cameron. Luke's dad, Larry, and I were old friends and we hung out together all weekend. It was a good time.

I considered myself to be a bronc rider, and I had ridden lots of bucking horses out on the range; but in truth my bronc riding was built on blood and guts and a lack of fear. I rode on strength but I possessed no finesse. I just stayed on. But after listening to the guys who were teaching the saddle bronc class, and their talk of lifting instead of pulling, staying back instead of leaning forward, of getting under your rein (whatever the heck was that supposed to mean!), I soon realized that I knew nothing about saddle bronc riding. It became apparent that had I tried to get Everett started in the craft of saddle bronc riding I would have assured his failure.

None of the students in the school could ride. Perhaps several of them got the hang of it quicker then Everett, although I really don't remember, but nobody was an instant star, and there was a continuous downpour of buck-offs. Everett was no different than the rest; he kept falling off over and over. I kept wondering if he would say he was finished and wanted to go home, but he didn't. One thing he seemed to get from the start was marking a horse out of the chute. He could do that with ease. Finally at the end of the third day, Everett got on an old black appaloosa that jumped about two inches off the ground, and he rode him all the way to the whistle. I was more excited than he was. We had us a bronc rider in the family!

When we got back home to the ranch I got a chute, which was nothing more than a four-sided steel box with a gate going out one side. There was no end gate so you had to load the bucking horse in the same gate that you opened to let him out. We tied this crude chute at the end of the roping arena.

The Babbitt Ranch had always had a problem with wild horses that came from the neighboring Navajo reservation to eat Babbitt grass. The standard procedure had always been to chase the Indian horses back across the fence, or the Little Colorado River, and onto the reservation. At times there might be several hundred of these unwanted Indian horses running on the Babbitt Ranch. There was always at least 20 or 30 because the supply never stopped. Now that I had a bronc rider in the family I started gathering these Indian horses, which for the most part were the equivalent to mustangs you might see on the Owyhee Desert in Nevada or the Red Desert in Wyoming; and I would drive them into the shipping corrals at Spiderweb. Mickey Byrne was a saddle bronc rider and was working at Babbitts, and he loaned Everett his bronc saddle and was on hand to give Everett free schooling, which was needed. Me and Clay or perhaps Mickey would rope and drag and push a wild horse, and stuff him sideways into the chute; and Mickey would help Everett get saddled. Clay and Jean Ann would run the gate, Mickey would flank, I would pick up, and we would have us a bronc riding session.

Almost all of those Navajo horses would buck a little bit, and occasionally one would really turn the crank, but they didn't have much power. They were absolutely perfect for a kid learning how to ride broncs. We usually would buck out two or three and occasionally four; and over the course of a year, we bucked lots of them. It was free and we had a blast. When we were done with them, I would haul them down the road 20 miles and dump them out onto the reservation, and usually within a week they would be back on the ranch.

At first there were lots of buck-offs even though the Navajo horses weren't big, powerful buckers, they could and would really throw some good wild-horse fits. Mickey Byrne was a good asset to have around at that time because he kept bringing Everett back to the basics. Mickey is not a big talker, but because of his knowledge and experience, what little he said was worth a lot. He sorta kept Everett going down the middle of the road. As each day of practice passed, I was reminded of what a disaster it would have been had I tried to get Everett started riding saddle broncs. The truth is I probably learned as much about riding broncs from Mickey as Everett did, but I just kept my mouth shut. I was smart enough to know that by keeping my mouth shut I wouldn't expose my ignorance.

We kept gathering more Navajo horses and Everett kept putting his bronc saddle on them. We were all having fun. It was a family event.

One day I gathered a bunch that had a big, good-looking buckskin gelding that I had not seen before. He stood over 15 hands and weighed at least 1150 and maybe more. He was definitely a standout among the rest of the Navajo broomtails. Everett got him saddled, and Clay opened the chute gate, and the horse came out bucking. He really took his head and bucked hard but didn't have a lot of kick just bucked straight with some altitude. Without a doubt he had more power than the other horses Everett had been riding. He stayed with him for five or six jumps, but then the horse knocked him out of the saddle and off the right side. Everett's foot stayed in the right stirrup just long enough for the buckskin to pick him up slightly as he came off the ground during the horse's next jump, and this drug Everett up and underneath the horse's body; and then he came loose, landing on his belly. When the horse came back down, one of his big feet landed right between Everett's shoulder blades. Everett's eyes sort of ran out on a stem, and he gasped for air. To me, from my perch on my pick-up horse, it looked bad; and it scared me to death.

I don't remember Everett making a lot of noise, no crying or wailing, maybe a little groaning; but I guess I got somewhat verbal with several outbursts like, "Are you all right? Are you all right?" Everett couldn't talk, but finally Jean Ann told me, "Shut up, he's all right." So I shut up, and Everett went to the house. The practice was over, at least for that day.

I kept wondering if he would stick with it. It takes a lot of buck-offs to get it figured out, and when a kid who is starting out does get one ridden it's usually ugly. Then one day we put a hame-headed bay Navajo mare in the chute. She wasn't real big, maybe a thousand pounds and 14.2 hands high. Her head was as big as her body. She was really ugly. When the gate came open, Everett had his spurs stuck in her neck, and she really bucked. About the second jump, he pulled his feet and spurred back to the cantle and then beat her back to the front end before she hit the ground. She bucked good for 10 seconds, and he got in time and spurred her every jump. He looked like Casey Tibbs. I was picking up and rode alongside of the big, ugly mare; and I was screaming like the big-bosomed girls on TV during the NFR at the Thomas and Mack Center. A friend had a video camera and filmed it, and after the first time I watched it, I had to shut the volume off because I was embarrassed at the amount of cheering I had done.

At Cameron, Arizona, north of Spiderweb 20 miles or so, there was an Indian named Randolph Beard. We knew Randolph well, and were friends with him. He was sort of a chief, or maybe bigshot would be a better description. Randolph was married to a white woman, and he owned lots of cattle that ran on the Little Colorado River up river from Cameron and bordering the Babbitt Ranch. Randolph had a grandson named Adrian that we were slightly acquainted with, and Adrian had a horse that he rode that was a good-looking grulla. One day I gathered Adrian's horse with a bunch of other Indian horses, and we put him in the bucking chute. I didn't recognize or remember the horse, but the kids did. It was probably Clay because he had a photographic memory for horses. But in spite of the fact that he was Adrian's pet, we loaded him in the chute, and Everett put his stock saddle on him. He was getting ready for the bronc riding at the Cowboy Reunion where everyone rode in their everyday stock saddle.

The grulla blew out of the chute and bucked like a scalded dog, and Everett spurred the heck out of him. We all laughed wondering if the horse would act different the next time Adrian rode him, if in fact he did ride him. The horse might have just run wild out on the Painted Desert the rest of his life, but we never knew.

Everett got to where he was staying on pretty regular and started placing in the high school rodeos his sophomore year. And that year, in October of 1992, I was offered the wagon boss job on the Diamond A Ranch at Seligman, and we moved over there. Richard Rudnick, one of the owners of the company that leased and operated the Diamond A Ranch, was trying to raise bucking horses; and they had a string of bucking horses at the ranch. I was told we could use them for Everett's practice. I found some old bucking chutes at an abandoned rodeo arena in Seligman, and I got them and set them up at an arena that I fixed up at a cow camp called Pica that was 17 miles out of town.

The Diamond A bucking string were bigger, stouter, heavier horses than Everett was used to practicing on, and for awhile the bronc riding was pretty ugly. On several occasions we drove home after a practice session were Everett had been bucked off hard multiple times, and the mood was pretty depressing. I wondered if, perhaps, he would give up. These new horses were a lot harder to ride than anything he had faced. Things were tough.

Somehow he found the intestinal fortitude to stay with it, and his junior year in high school he won first on just about every horse he got on. At the state finals he rode a little blue bronc that Dennis Reiners owned. Dennis Reiners was the world champion saddle bronc rider in 1970. The blue horse was named Fuzz Nuts, and Everett won the third go-round at the state finals and cinched the bronc riding title for the year. It was a proud moment for all of us.

With Everett having all the fun and being in the spotlight after his enrollment in the Lyle Sanky rodeo school, in January of 1992, Clay decided he would follow in his brother's footsteps; and he enrolled in the same school a year later. He was 14 years old and had broke several horses to ride and had a little bucking-horse experience, so he was ready to go the same route as his brother. Everett was there alongside of him and full of excitement and advice. After all, he was now a veteran. Clay showed just as much ability and enthusiasm as any of the students for the first day, but as the second day of the school advanced he seemed to lose interest. It wasn't that he was scared or worried, but, instead, it was obvious that he was thinking about the situation, evaluating the possibilities.

After running a full section of broncs through, with every student getting on one and then having a little discussion about what each individual needed to work on, the instructors ordered that another pen of broncs be run into the chutes. "Get your saddle on one, boys," an instructor said.

Clay sat over against the fence behind the chutes doing nothing, so Everett said, "Come on, Clay, get your saddle. Let's get going."

Clay looked at him as if he had been pondering the mysteries of the universe, or perhaps Einstein's theory of relativity or maybe something as simple as how to gather an ornery cow, and he told Everett, "You know, if we rode these hame-headed son of a guns every day for a year, none of them would ever amount to anything!" And with that little piece of wisdom said, Clay's career as a saddle bronc rider was over. He was his own man and would make his mark in other ways.

There was a really rotten horse at the Diamond A Ranch named Dill that would buck people off and try to kick them in the head or some other trick trying to hurt someone. A cowboy named Randy Rutledge rode him and got along with him fine, but he would keep him ridden down and tired, and he wasn't afraid of him. Several other good hands rode him, but nobody liked him; so I put him in the bucking string, and one afternoon during a practice session Everett put his saddle on him. The horse blew out of the chute jumping high with no part of his body closer than four feet from the ground. He was rank. About five jumps out of the chute, the horse turned his neck around in a U-shape and stared at Everett with both of his eyes. They were red. I swear they were. When he did this, it threw a bunch of slack into the bronc rein, which any bronc rider knows is a difficult situation to overcome; but Everett lifted his left hand holding the rein high and way out in front of his chest and got under his rein and sank deeper into his saddle. That was a move that only a pro could have negotiated. Watching that was a defining moment for me. I thought to myself, "This durned kid is making a bronc rider!" Several jumps later Dill bucked Everett off, but I felt that he had reached a higher level.

In Everett's junior year in high school, he won first in 20 of 24 go-rounds at the high school rodeos. His senior year he was injured and missed several rodeos but had almost as good of a record. When he turned 18 he bought a PRCA permit and won first in the first pro rodeo he ever entered. I'm glad I didn't teach him to ride broncs.

Chapter Sixteen

One morning when Clay was about eight years old, I took him with me to help me move some cows and calves out of a small holding pasture and up into 89 pasture, about four or five miles. We had 40 or 50 cows with babies. It was one of the first times I took Clay with me for any distance. I told him to just follow the drags, and I would take care of pointing the herd of cattle the direction they needed to go. I told him it was important to not push them hard but to give them the room to stay paired up and strung out. We got them going, and the leaders took off and went to traveling. After awhile I loped quite a ways ahead to turn the leaders one way or the other, and Clay was left alone for quite a spell. When I went back to the drags, Clay was going along fine, and the cattle all seemed to be happy and walking along. I asked him if he knew which calf belonged to which cow. There were probably 30 pair in close proximity to where we were. He said, "Sure," acting as if he didn't even need to think about it.

So I challenged him, "Okay, tell me which cow belongs to which calf." Without pausing for even a second, he went to pointing out which cow and calf belonged to the other. Some of them were walking close to each other but some were not, and without fail he told me what the pairs were. I had not coached him or warned him that I was going to ask him about it.

On the day that Jean Ann and Everett went to Dawn McFarland and Justin Salcito's wedding, Clay stayed home with me, and I took him with me out onto the Navajo reservation north of the lava beds that come out of SP Crater. We went way out north toward Gordy Dam looking for six yearling heifers that someone had seen running out there. It was a cold, windy October day. I was riding a good horse I had named Jerky, and Clay was riding Beaver. It was a few days before Clay's 12th birthday.

After a long ride and much trotting and loping we located the six heifers and started trailing them back south toward a gate going into 89 Pasture on the east edge of the lava beds. At one time there had been a cattle guard on the spot but it had filled in with sand and was fenced over, and there was a wire gate at the spot. There was also several old abandoned car bodies on the reservation side of the fence and numerous other pieces of discarded appliances and stuff typical of what you might see on Indian reservations. There was also numerous pieces of garbage such as plastic bags or feed sacks and discarded clothing laying around, all of which was fluttering in the wind in the vicinity of the wire gate.

I loped up and opened the gate, and we drove the six heifers up to it and tried to put them through, but they refused to go anywhere near it, but acted as if all of the trash and paraphernalia was spooking them. We got them as close as we could and held them and waited. They watched and we waited, they would spook at a sack ruffling in the wind and try and run off. We would drive them back and hold them. We drove them up and down the fence going east to west and then west to east, and then held them some more. I tried to act like a good old cowman who had patience, but the heifers wouldn't go near the gate. I almost thought they were laughing at me and Clay.

The wind was getting colder and the shadows longer, and finally after much waiting, my patience abandoned me. I took down my rope and ran and roped a heifer and drug her through the gate. After getting her on the Babbitt side of the gate, I hollered at Clay who was holding the remaining five heifers. "Come and heel this heifer," I said.

"I don't have a rope," he hollered back.

"What do you mean, you don't have a rope?"

"I forgot it. It's at home," he answered.

I jerked the heifer down and got my rope off. They were all weighing about 650 pounds. I then went and roped another and then another, jerking each one down and getting my rope off. Clay held the others as close at hand as he could while whipping and spurring Beaver to get the job done. After getting three of them on the Babbitt side of the gate, I rode up to him and handed him my rope. "Here, you rope the rest of them," I said.

Without answering he took after a heifer and got a loop on her and got her onto Babbitt territory. He got her wrapped up in the rope and pulled her down and then dismounted Beaver and ran down the rope while holding onto the tail of the rope but keeping a dally on his horn. About the time he got to the heifer, she jumped up to her feet, and Clay ran back to Beaver and stood by him and hollered at me. "What do I do now?" he asked.

Through the wind I said, "I don't know, but you need to figure it out because you're got two more to catch, and it's getting late."

He figured it out and caught the remaining two heifers and got them through the gate and then handed me my rope back. Nothing more was said about it. As we traveled toward home, I could tell Clay had a glimmer in his eye. He might have acted like a gunsel and forgot his rope, but he had also made a hand and never missed a loop, or lost his rope, or broke a leg on a heifer.

Going to the Arizona Cowpunchers Reunion and Rodeo was a big deal to all of us when Everett and Clay were growing up. The summer Everett was turning 13 and Clay was going on 12, we kept up a bunch of horses in the corral at Spiderweb. The bunch included the horses that Bill, Vic, myself, the boys, and all the crew were going to be competing on; plus there were several others in the bunch. There is a very large waterlot on the north side of the corrals with two 1000-gallon steel water troughs located in a corner of the big lot. The two troughs being located several hundred yards from the gate going into the horse corral. A 150 yards west of the water troughs, there is a big gate leading into the horse pasture which is three miles long.

Early in the morning and way before daylight, I told Everett and Clay to catch their horses and saddle them and drive all of the horses that we had kept up out into the big lot and let them drink at the water troughs. We were going to leave and head to the rodeo not long after sunup. The boys caught their horses, Everett catching Waddie and Clay catching Charro, and they saddled them. These two horses were half-brothers out of the same mare. She had twelve horse colts in a row, six out of one stud and then six more out of a different stud. They were all good horses. After getting saddled the two boys stepped on their horses without a bridle, but, instead, they left their halters on Waddie and Charro and then opened the horse corral gate. Putting bridles on their horses would not have been as cool as riding with halters, or so they thought. The plan was for one of them to ride over to the big gate several hundred yards away and sit in the gate and not let the horses escape out into the big pasture. The smart thing to do would have been to trot out there first and shut the gate, but that would not have been as cool.

The fact is the horse corral gate was open, and the horse herd was turned out, and Everett and Clay were riding with halters. The horses went through the corral gate and on out to the water troughs and then turned toward the horse pasture gate. The bridleless horses were feeling frisky and came untrained, and the race was on. The whole herd was running full blast when they cleared the horse pasture gate, and Everett and Clay were hanging on for dear life being completely and totally out of control. They stampeded to the northwest for a half mile or so, and finally Clay was able to use his hat to cover Charro's eye and steer him; and somehow they turned the stampeding herd back to the horse corral. The entire wreck was done at out-of-control speed, in pitch black darkness, running through badger holes and washouts, but somehow nobody got hurt. Everett and Clay did not tell me the story until 20 years later, and at the time I was totally oblivious that anything of the kind had taken place.

In the fall of 1992, I was offered a job running the Diamond A wagon, which is the same as saying the foreman's job, on the biggest ranch in Arizona. It was hard to leave Babbitts because we were comfortable there. We liked the people there and they liked us, and in spite of the fact that I had a history of changing jobs too often, we had managed to stay there for seven years. Actually, I had worked there a total of 15 years. It was more of a home to me than anywhere else I had ever been, but I took the job at Seligman, and we moved. The house we moved into was in the middle of town and had always been the manager's house, and that was what the outfit provided for us. Jean Ann hated it and never quit hating it until the day we left three and a half years later. She was not a town girl. I was gone all of the time chasing 10,000 cows across close to 1200 square miles, so where we lived didn't matter to me.

One of the agreements I had set in concrete when I took the job was that I could work my two sons, at full cowboy wages, all I wanted to; and work them I did. Everett had just turned 15, and Clay turned 14 several days after we moved in.

The ranch had a big crew of 12 or 15 men all of the time. At one point in June of 1993, I had 22 men on my crew including a cook and horse wrangler. Everett and Clay had been around crews of cowboys before and had a good fundamental knowledge of working cattle in situations like they were now exposed to on the Diamond A, but the mental pressure of being thrown into new territory and strange people was a challenge to them. The cattle were extremely spoiled and hard to handle, the horses would kick and fall over backwards or try to buck you off, and there was a considerable amount of alcohol consumption among certain members of the crew. There was probably some drugs also, but that was kept well hidden from me.

On one of the first days we worked there, I led the crew out west of Rose Well into a pasture called Owen Dam. I kept Clay close to me to flank for me. Quite awhile after the men had been scattered out through the draws and ridges and cedar trees, I rode up on a high point where I could see for miles and soon Clay caught up to me, and we sat and let our horses blow for awhile. I looked across several miles of country and could see, a long way off, a big bunch of cattle turning back the wrong way and trying to escape, and it looked like they were going to be successful. I told Clay, "Lope over there and tell Randy to relay the message to the next man that a big bunch of cows are turning back up that draw over there. If someone doesn't hustle up there, we are going to lose them."

So Clay took off, and in several minutes he reached Randy who was about a quarter of a mile away to the south of where Clay and I had been talking. "My dad said to tell you that there is a big bunch of cows getting away and running from the drive up that draw. If someone doesn't get to them, we are going to lose them."

"Let me tell you something, kid," Randy said, "we all work for your dad but that doesn't mean we have to take orders from you. Do you understand?"

Clay answered in the affirmative and then rode back and positioned himself halfway between me and Randy. The cows got away, and I never heard what Randy had told Clay until 25 years later.

On one occasion we had three or four hundred head of cows plus calves and bulls in the shipping corrals at Keseha on the Diamond A Ranch, and I was trying to sort them. I wanted to cut some cows through a gate that was being watched by a buckaroo who was not a good hand, even though he was very opinionated and constantly making comments about how good or bad a job everyone else was doing. Actually, he was the worst person on earth to try and cut cattle by, and he had a very caustic personality. Everett and Clay were close by as well as several other cowboys who were trying to hold a line on the herd of cattle. Things weren't going well. The buckaroo was supposed to be positioned in the gate, and when I got a cow cut through the gate and into a different corral, he was supposed to keep them from escaping and coming back through the gate and into the herd that they had just been cut out of; but every time I would head toward the gate with a new cow, the buckaroo would abandon his post and ride into the herd to help me, and the cattle that I had already cut would escape and return to the herd they had just been cut out of. We were not making any headway. The buckaroo was making lots of critical comments.

While all of this was going on, Jim Fancher and Cody Cochran rode up and were watching from the outside of the corral. I had sent them off to do something else, and having finished what they had been told to do, they rode up and were watching the wreck that was taking place.

After about the fourth time the buckaroo let the cattle I had cut out get back into where they had been cut out of, I sort of short circuited, or more correctly said, I got mad. I looked at Everett and Clay and in a heated voice said, "You guys are going to have to help me!" My tone of voice insinuated that they hadn't been helping me, but, actually, nothing bad that had happened was their fault; so when I realized that my angry comment toward them was unwarranted, I got madder. The buckaroo started to make a sarcastic comment, and I told him, "If you would get your fat butt in gear and watch that gate instead of running into the herd where you don't belong we would all be better off." Then I looked down on the far side of Clay at a gunsel kid who had no idea anything was amiss, and I hollered at him and said, "And if you turn your horse's butt to a cow one more time today I'm going to make you wrangle horses afoot for a month. Do you understand?" The kid didn't have a clue what I was talking about. Then I looked at Cody Cochran and Jim Fancher, who were both top hands but out of the corral and therefore unable to help me. Cody was laughing, but Jim, when I looked at him, turned his horse and rode off a hundred yards and stepped off behind a cedar tree and dropped his pants in an obvious attempt to do some private business.

I went back to cutting cattle, and we eventually got something accomplished. The next day, Everett, who was 16 years old, said to Jim Fancher, "Boy, you were lucky to have to go to the bathroom yesterday when we were having all that trouble sorting those cows. That was good timing."

"I didn't have to go; I just rode away and dropped my drawers because I was afraid your dad was going to have me help sort those cows."

In the spring of 1993, I camped the wagon at Number 2 about six miles north of Rose Well, and we worked out of that place for a week or so. It was in early April and was pretty chilly in the morning. Everett had a four-year-old in his string named Sneaky Pete who was a brown colt and a good horse. Sneaky Pete had been turned out all winter so he was fresh and fat. We caught horses one morning when it was just light enough to see, and I caught Sneaky Pete first so Everett could get him saddled because I knew the horse was fresh and because of that he might be a little feisty. About the time I got the last man's horse roped, Everett had Sneaky Pete saddled; and he led him out a hundred yards or so away from everyone else, and he proceeded to get on him while everyone else was getting saddled.

When Everett got on, Sneaky Pete blew up and went to bucking and headed toward a fence 20 or 30 yards away. The horse was squealing and bawling and bucking hard. When he got to the barbwire fence the horse turned to the right and toward a line of men who were in various stages of getting their horses saddled. Sneaky Pete was really bucking now, and Everett was really spurring him, and the two of them went down the line scattering men and horses and saddles. The wreck was on. Just before Everett and the bucking bronc reached a fence corner, they came to Jim Marler. Jim was just trying to throw his saddle on a horse when Sneaky Pete came jumping toward him. Jim cursed, his horse jerked away from him, and the saddle that he was trying to throw over his horse's back flew through the air. Sneaky Pete almost mowed Jim down, but he jumped, just in time, and went over the barbwire fence. The top wire of the fence caught on Jim's belt, and he teetered for a second like a seesaw, but then went toppling over to the other side. As Sneaky

Pete bucked past him a few inches away, Jim's high-heeled boots were in midair and his black cowboy hat was touching the ground. He was upside down. It was hilarious! Jim didn't see any humor in it at all. But I did. Sneaky Pete eventually made a durn good horse.

The cattle on the Diamond A Ranch were very spoiled in the early nineties. A good tally had not been taken on the cowherd in 20 years, and there were unbranded cattle all over the ranch. Crews of men had been making big drives with too much room in between cowboys, and the cattle had learned to hide when they heard cowboys approaching. In the trees cattle would actually go toward the noise and then sneak between men and run off going the opposite direction. We made adjustments to the situation and changed a good many things and eventually things got better. We gathered the cattle and got a count on the cows and found that the outfit had 500 more cows than the owners knew they had. Harvey Dietrich, who was one of two partners we worked for, told me this on two separate occasions. Everett and Clay were a very important part of me being able to do my job as wagon boss on the Diamond A Ranch. They worked for me and made a hand.

One evening not long before sundown some cows came drifting into Rose Well from the Rogers Pasture. We were down to a small number of remnant in that pasture so we were wanting to get every cow we could find. Clay and another cowboy saw the cattle trying to sneak in without being noticed. The cattle were like that, they were sneaky. Clay and the other man saddled up and went out and tried to pen the cattle, about four head; but when the cows saw them, they turned tail and proceeded to run off. Clay had two piggin' strings and used one on each of two cows to tie them down, and then he roped a third cow and tied her down using his red wild-rag scarf. The other cowboy caught the fourth cow, and so they got all four of the cows securely locked in the corral and therefore accounted for.

There was never an end of something to do at the Diamond A Ranch. There was over 10,000 cows on the outfit at the time. I still have my tally books from when I ran the outfit, and I could show them to you to prove it. We would work for 90 days in the spring and 90 days in the fall, without taking any time off, just to get the main part of the work done. The other 180 days of the year we gathered and worked the cattle we had missed when roundup was going on. All we did was saddle a horse and ride every working day, doing something with cattle somewhere on the ranch.

Chapter Sixteen

I didn't play a lot of baseball or other sports with my kids. We didn't go on lots of spectacular vacations. A great deal of the time when my kids were growing up, we were barely scraping by on cowboy wages, and we drove old used pickups and lived in company housing; but I did teach them how to work and make a hand. We spent a lot of time together, and when they were 17 they were better cowboys than I was when I was 22. They knew how to conduct themselves around a crew of men, and they weren't lazy or drunk or liars. When they left home they knew how to work and accomplish something.

Chapter Seventeen

In the summer of 1976, I stayed in the bunkhouse at Cedar Ranch and broke some horses to ride. Harvey and Janet Howell were living in the big house, which was downhill a few feet from the bunkhouse where I and the old, one-eyed waterman, Mike Lenton, lived. I always liked Cedar Ranch, especially in the summer when it would rain, you could look out to the north for a hundred miles and watch the storms as they blackened the sky, and then black columns of rainfall would be born, sometimes a mile away or perhaps 50 miles; it didn't matter, whenever the rain fell it was welcome; and from a vantage point at Cedar Ranch, the variety of different landscapes were limitless. From where you would be seated there were cedars, pinion pine, and ponderosa pine stretching northward; and dropping in elevation, the black malpai rock changed to white limestone and then the black rock would reappear and then turn to red sandstone. Mesa Butte rose up out of the limestone rimrock at the Tubs four miles away, and 20 miles farther big Gray Mountain with its limestone cliffs tried but failed to hide the Painted Desert far beyond that. I liked it there.

Jean Ann and I were going together. She had found out my family were Democrats, and she had forgiven me. She was working at forgiving them. She thought I was somebody. I knew I was somebody.

There were no cross fences in the Slate Lake allotment in those days, just one big pasture. The way I remember it, it was about a township in size with lots of elevation changes and evergreen trees.

One day in July, I rode out of camp on a flaxy-maned sorrel gelding that was three years old that I had named Blondie. I had about 10 rides on him, that day making number 11. I rode west toward Babbitt Lake, and after a mile or so, I veered off of the road on the north side. I dismounted and cinched up my saddle real tight and stepped back on and took down my nylon rope that hung on a string from the fork of my saddle and tied it off hard-and-fast to the saddle horn. There were several cow and calf pairs grazing nearby, and I planned on roping one of the calves around the neck. The calf was healthy and did not need to be caught, but my colt was green and needed to be trained. Besides I was a roper and ropers are addicted to roping, no different than alcoholics are addicted to whiskey.

I pursued the calf, and almost immediately the chase turned into a high-speed one. This did not surprise me, but, on the contrary, it would have surprised me had it not been high speed. That would have been a conundrum as a slow calf would have been sick and need to be caught and doctored, whereas a healthy calf would run fast. I needed the excitement, and my horse needed the experience, and the calf was very healthy.

There were lots of pinion trees and cedar trees everywhere; there were outcropping of black malpai rock, some six inches high and others several feet high. They were everywhere, and as far as you could ride it was the same.

We ran fast, and soon Blondie was upon the fast calf; and I only needed another 10 feet, and I would catch him for sure. I had done this type of thing countless times before. I was good at it.

I gauged my need to close the gap another 10 feet, which would only take several seconds; and I saw a very large cedar tree looming ahead in our path. I knew that within 40 feet after passing that tree I would catch the calf who now ran straight toward the center of the tree that had full branches reaching all the way to the ground. The calf, a heifer weighing 300 pounds, ran up close and quickly darted around the left side of the tree.

Blondie, being a split second behind the calf, had his ears laid back against his neck. I suppose I leaned slightly to the left, thinking Blondie would follow the calf around the left side of the tree. He ran close, almost crashing into the tree, and quickly, to my surprise, ducked to the right and around that side of the tree. The quick unexpected move to the right unseated me, and I fell out of the saddle. We were moving at a blistering pace and both of my hands were near ground level as the rocks whizzed past, and my head was down there also. My left knee was bent, and my left foot was still in the stirrup. My right leg was hooked around the cantle of my saddle, and the inside of my knee and the calf of my right leg gripped the cantle. At least 90 percent of my body weight was off center on the left side. I was dragging, only I wasn't dragging. I was in limbo, neither riding nor dragging, and like that we raced along.

I reasoned with myself, thinking faster than the speed of light. I told myself to just turn loose of all of my holds and drop to the ground. I told myself that no colt with 10 rides was going to put up with this but would blow up and buck and throw me down. Holding on made no sense to me with my cowboy hat brushing against grass and rocks that we raced over. But I didn't just fall off. I didn't give up. I pulled with my right leg against the cantle and somehow I lifted my right arm up and my hand grasped the saddle horn. Somehow I righted myself, and there in front of me was the calf still running at high speed.

When I was up, straight in the saddle, I looked down and saw that the loop of my rope had somehow fallen down and drew up around my waist. I slowed Blondie down and realized that if I had fallen to the ground I most certainly would have been drug to my death.

But I wasn't.

August 19, 2016 was a nice sunny day as we drove away from the 4 3 Ranch, 40 some miles north of Lusk, Wyoming going south toward Cochise County, Arizona. Jean Ann and I had made the trip north from the ranch were we had lived for 19 years at Apache, in the far southeastern corner of Arizona, to visit Clay and his wife, Lexie, and their two sons, Miles and Grant. We had enjoyed a very nice visit with Clay and his family. I had got to ride with Clay and Miles and the 4 3 crew and help them brand calves for several days.

We had made this trip many times and had probably taken every route possible on highways in Arizona, New Mexico, Colorado, Utah, Nebraska, and Wyoming; but the fastest route is right down I-25 to Hatch, New Mexico where we turned west toward home, 65 miles southwest of Lordsburg, New Mexico. Usually we would stop somewhere halfway, like Trinidad or Raton; but on this day we kept driving, always thinking we would stop when we got to the next town.

Chapter Seventeen

We reached the north side of Albuquerque about sundown, and I was driving while Jean Ann rested. I planned on stopping and getting a room, but every time I saw a decent looking motel I was past the exit before I saw the motel so we kept driving. It was about half dark when we reached the interchange of I-25 and I-40, and Jean Ann was awake, and we talked about what to do. I had driven a pickup pulling a trailer and rope horse many times from the fairgrounds in downtown Albuquerque going to and from roping events, and I could make it home in six hours. If we kept driving we could be sleeping in our own bed by one in the morning. We decided to keep going. Jean Ann was fresh and said she could drive for several hours, so the decision was made; we were going to stop and switch drivers somewhere on the south side of town.

It was a Friday night and by the time we reached exit 209, which goes to Isleta Pueblo, it was completely dark. There was very little traffic either direction. I pulled off and drove up the long off-ramp to the top of the hill where the bridge goes over the freeway. I stopped at the stop sign and then proceeded downhill on the on-ramp and went several hundred yards, or halfway to where the ramp merged with the southbound lane of the freeway. I pulled off the side of the pavement and stopped, putting the car, which was a brand new Toyota Corolla painted bright red, in park. I opened my door but sat there exhausted.

Jean Ann opened her door and jumped out and quickly ran around in front of the car with our headlights shining on her smiling face. There is a deserted-appearing dirt road going west from the stop sign at the top of the hill, which was now directly behind us several hundred yards. As Jean Ann ran around the front of our car, I could hear a car coming up the dirt road from the west. About the time Jean Ann reached the front of my driver's side door, which was open, the car slowed down for a second and then turned south toward us, and its lights shone bright in the rear view mirror. Jean Ann was standing by the front fender in front of the door waiting for me to get out so she could get in. I reached my hand out and took ahold of the door handle and turned slightly and proceeded to get out of the car. The car approaching from behind was accelerating fast and closing the gap between us, and I decided to let him pass before I stepped out. That decision saved my life. I sat there with my head turned slightly to the outside and pulled my hand back inside the car. I was aware of Jean Ann standing directly in front of my door. The approaching car's engine was screaming, its RPMs racing higher with every approaching foot. It got louder and louder, and the lights got brighter, and the engine was sounding like an approaching F-16 jet . . . .

Suddenly, in an instant, even faster than a lightning strike, my car door exploded. It was bent around and crashed into the front fender. Glass was flying through the air with a weird tinkling sound, like a million wine glasses falling onto a tile floor, and it sparkled in bright effervescent splatterings, each of which reflected headlights in prisms of horrific particles of a dream. The sound of sheet metal being twisted and sheered and crashing sank into my brain at the speed of light. At no time through any of this did the sound of our assailant's engine's RPMs cease to climb, but the car, without a falter, roared on; and when it reached the southbound lane it was probably doing 90 miles per hour, racing on southward toward somewhere.

From the time the devil hit our door until he was on the freeway was probably less than two seconds. I barely had time to straighten my head and look forward through the flying glass, and then I saw it. I saw Jean Ann's head bouncing down the pavement like a basketball. It had been cut off at her shoulders in the crash and was bouncing down the road. I could see it in my headlights. Her torso and arms and legs had been separated in the violence, and she was in pieces bouncing, bouncing; and I saw them come between me and the taillights of our enemy as he entered the southbound lane.

Sometime, about the time the other driver entered fully into the southbound lane, I started groaning, mourning, grieving, "Jean Ann, I've killed you! Jean Ann, Jean Ann, I've killed you!" I had watched her head bounce down the road. "Jean Ann, I've killed you!" I knew at that moment what someone in the World Trade Center felt seeing the airplane flying directly into their window. I knew at that moment what someone feels when a terrorist's pipe bomb explodes and scatters hundreds of nails a few feet away, killing people close enough to touch. "Jean Ann, I've killed you!"

I sat there, was it 10 seconds? Was it a minute? I don't know, I was utterly hopeless. I was confused. I was watching body parts bouncing down the pavement. How long does it take a person to reach rock bottom? Not long.

"Ed, I'm over here." My eyes bugged out.

"Ed, I'm here." Jean Ann's voice came to me very faintly, "Ed, I'm over here." I couldn't see her. She had been directly in front of my door. Her body had exploded into pieces. I stepped out of the car that was still idling and went around behind, and up its right side. There she was. She was laying on her back, her head resting on the pavement about three feet in front of the right headlight. Her hands were by her sides, and she was as straight as a board, out in front but perpendicular with the front of the car. She stared up as if in a trance. She was beautiful. Her hair was brushed perfectly and was laying down behind her back as if someone had held her head and hair perfectly and gently laid her there. It was equally divided onto both sides of her head and shoulders. There wasn't a scratch on her. There was no blood. There was no broken glass mixed in her hair or laying on her clothes. She looked at peace staring upward. She didn't seem to see me.

I knelt down and started to lift Jean Ann up and into my arms and her expression changed, "Don't pick me up. I'm kind of hurt," she said.

"Are you hurt?" I asked.

"Yes, when I move my knee hurts."

"I'll call 911 and get some help," I said.

Just then a car pulled up and stopped, and a young woman rolled down her window. "Are you all right?" she asked. A young man sat behind the steering wheel. They looked us over with genuine concern. "We will call an ambulance for you," the young woman said, and she began dialing on her cell phone. They waited by us until authorities began to arrive, which seemed like only minutes.

Soon several Bernalillo County deputies were on the scene and not long after that an ambulance. Jean Ann explained to the paramedics that her knee hurt. And after the paramedic pushed and prodded on her, it was decided that her hip or pelvis might be hurt also.

The deputies asked me what had happened so I told the story to them. One deputy walked down the pavement looking at the car parts that had scattered on the asphalt. He came back and told me that he had found a piece of plastic from a bumper with writing on it that said the vehicle that had hit us was a late model black Honda Civic. That was the only evidence that was ever found.

I walked around the front of our vehicle and saw a dent in the fender, with the imprint from her Wrangler's pants' pocket in it, just in front of the driver's door exactly where Jean Ann was standing when the black Honda hit the car door mere inches away from where she had been standing.

The emergency people loaded Jean Ann into the ambulance and told me that I could ride in front of the ambulance on the passenger side while the paramedic rode in back with Jean Ann. We took off driving north to a hospital in Albuquerque a dozen miles away.

I used my cell phone and started calling people and asking them to pray for Jean Ann. I told them about the accident and that she was injured, but I had no idea how bad. I told them that there was no blood or visible damage, but she had pain in her pelvis and knee.

I called Everett, who was in Capitan, New Mexico with his family where he was competing in a rodeo. He said he would leave immediately and come to Albuquerque to be with me. About three hours later he arrived at the hospital, having been driven there by several close friends, Bobby Valdez and Clayton Smith.

At the hospital they took multiple x-rays of Jean Ann's body that showed she had several hairline fractures in her pelvis and one on the side of her right knee. The emergency room doctor said that her pelvis would heal in time and probably be fine, but there wasn't much that could be done except to wear an elastic girdle type bandage for an undetermined length of time and take lots of pain medication. He said that her knee might require surgery, but an orthopedic surgeon would have to make that decision later. He also said that he guessed that Jean Ann was going to have to stay in Albuquerque for several weeks and go through some physical therapy.

They kept Jean Ann in a room in the emergency area of the hospital but said that she would probably be admitted into a regular room the next day. She was very tired looking and under some heavy-duty pain medication. Around two in the morning, Everett and I called a taxi and went several miles away and got a room and tried to go to sleep. My head was spinning. I was exhausted and worried about Jean Ann. I wanted to get home. I wanted to get her home. I don't know what I would have done if I hadn't had Everett there to help me.

Early the next morning we got another taxi and went back to the hospital. They were talking about doing surgery on Jean Ann's knee. They were talking about officially admitting her into a regular room, but couldn't decide. Maybe if they didn't operate on her knee they could discharge her straight into a physical therapy facility. There was a lot of confusion. In spite of the confusion, the nurses who were taking care of Jean Ann were great and the doctors were great, just undecided about what to do. I sat around and worried.

Everett went down to a Hertz car rental place at the Albuquerque airport and rented a minivan so we had some wheels. Then he found out from the police where our wrecked car had been taken and went there to get a few of our personal things that had been left in the car. While he was there, he looked the car over real good. It was a brand new car and very shiny. He saw the same dent in the left fender a few inches in front of the door's hinges that I had seen, and there, as plain as day, where scratches made by the rivets in Jean Ann's hip pockets as well as the W of the Wrangler logo imprinted into the red paint by the thread sown into the pocket. The door smashed her hard enough to use her hind end to make these marks.

While in the hospital, I became more stiff and sore as my muscles responded to the abuse they had been put through. On the first day, an operation on my knee was advised, but we said we did not want to have an operation, so that moved us forward toward the rehabilitation people needing to come and test me to see what I could do and what I could not do.

Late in the afternoon they came into my room and put me through their paces. They showed me how to use my good leg to move my broken leg over so I could get myself out of bed. I managed to move to the edge of the bed and stand weakly by it holding on to a walker. Alarms were blaring about my oxygen level. The alarm wasn't needed because it was obvious to all that I was exhausted from that small amount of exertion. They helped me get back into bed and said that I would have to go to rehab and there recover enough to pass the criteria required for being allowed to go home. The two women listed the required accomplishments: get out of bed, walk with the walker to a chair, sit down, use their tools to rope my foot and get a sock on it, and walk back to the bed and get myself back into it; all with no help.

That night, whenever I woke up, I tried to exercise to get my strength back because Ed had made it plain that he was not going home without me. I knew he needed to get home and get to work, and we didn't need the expense of a motel room for who knew how long. The exercising made me get the chills and start shivering, so I would then call the nurse, ask for another blanket, and drift back to sleep, only to repeat the processes again and again throughout the night.

The next morning, Everett came into my room before Ed. I felt terrible and must have looked terrible. I told Everett, I'm going to have to go to rehab. I can't go home. He looked at me and agreed. He came to the bedside and took my hand and prayed a simple prayer for God to give us strength and wisdom. I drifted back to sleep.

When I next opened my eyes, Ed was standing at the foot of my bed. I felt like a new penny. I must have looked like a new penny. I told Ed, "Let's get the nurse to call the rehab ladies and see if I can pass that test." He agreed. We called the nurse and made our request. I am not sure if she was the one that had brought me blankets all night, but she sure looked skeptical about me passing the test. But there was also a hint of perplexity added to her doubtful expression as she looked at me. She agreed to call them but told us that the rehab people did not come in on Sundays. She left and was back in a minute, more than perplexed; she was amazed: The rehab people were coming to the hospital this Sunday, and they would come to see me.

In no time at all the rehab women looked around the curtain surrounding my bed with total skepticism. Ed and I watched as their expressions changed to one of wonder at the sight of me. "You're feeling better," one of them ventured to say.

"Yes I am. Would you give me that test so I can be released and go home?" Still harboring some unbelief, they agreed to watch me as I went through the drill. Over to the side of the bed I scooted, then off the bed. With hands firmly on the walker handles, I marched to the chair (two steps), sat down, roped my foot. By now the women were standing back as if to get too close to me would put them too close to something they were afraid of.

I got myself back beside the bed, but I couldn't quite get up into it. They said that was fine. They were not going to come between me, what was happening to me, and what I wanted—to go home.

After getting checked out of the hospital, Ed and Everett got me loaded into the back of a rented van on a cowboy bed donated to the situation by the men who had come to get Everett to take him home when it had seemed like that was how it was going to have to be: Everett going home, and Ed and me staying in Albuquerque. Everett drove and Ed sat in the passenger seat. I drifted off to sleep, woke up, and looked up at Ed turned around looking at me, smiling, asking if I was doing okay. That same scenario happened all the way from Albuquerque, New Mexico to our home south of Apache, Arizona. Even in my weakened condition and not up to my usual analytical skills, I wondered if Ed had ridden turned around backwards the entire trip or did he have a sixth sense to know when I was going to wake up. Either way, I was convinced that he was very happy to be getting me back home.

So what had happened between the moment the door crushed Jean Ann's body into the fender and the split second later when I turned and saw car parts bouncing down the highway illuminated by our headlights, making me think I was seeing her head and other body parts bouncing down the road? I had been aware of Jean Ann standing in front of the door looking toward me. I heard no screaming or any sound of any kind coming from Jean Ann. When I found her she was laying back on the pavement with her head several feet in front of the right headlight, I wondered how she got over the top of the engine. Why wasn't she cut up and bleeding? Why wasn't there glass in her hair and on her cloths? Her hair was combed perfectly. How could she have moved that distance without me witnessing at least the tail end of the movement?

Jean Ann's memories, and they are very vivid, are that she felt no pain when the collision took place. She remembers being, as if suspended, above the hood of the car and seeing millions of lights that looked like one might imagine stars to look if one was much closer to them. She says that she never in her life felt so much peace and tranquility. Everything was perfect. She wondered to herself if the lights were pieces of broken glass that came from the car window, but she didn't think so. There was no glass on her or around where she lay. She and I have talked this over many times and have relived it in our minds, and we are convinced that God's presence came and He wrapped Himself around her and kept her safe and laid her down gently on the pavement in front of the car.

I think me calling out in grief that I had killed her, brought her back to earth. To awareness. And she thought she better call out to me so I wouldn't put the car in drive and then run over her because I had decided to chase down the other car and beat up the driver. I actually never thought of going anywhere. I was completely numb with terror and grief. At that time I had no plan for the future.

When we got home, Jean Ann was confined to bed quite a bit of the time but got better every day. A neighbor girl, Iliana Smith, brought lots of good food for us to eat and helped clean our house. People were good to us. We were advised to go to doctors for surgery on the bad knee and physical therapy. We did neither, but, instead, we prayed; and within a year Jean Ann was running around like someone half her age. She is whole.

From the very start of that whole tragic wreck, even as early as the ambulance ride to the hospital in Albuquerque, I knew the authorities would never apprehend the person or persons who deliberately tried to kill us. There was too little evidence, and the authorities have way too many crimes to solve. I soon realized that it was better that way. There was no messy trial. There is no face to hate. Those are things I don't need. I need my family. I have Jean Ann and she is whole. I have Everett and Clay and their wives, Leah and Lexie. We have a bunch of superstar grandchildren, and we all love each other. We talk and we forgive each other of our sins. And that's enough. Amen and Amen . . . . . .

Chapter Seventeen

www.ingramcontent.com/pod-product-compliance
Lightning Source LLC
Chambersburg PA
CBHW081920170426
43200CB00014B/2775